# WHAT IN THE WORLD

# ARE X-RAYS?

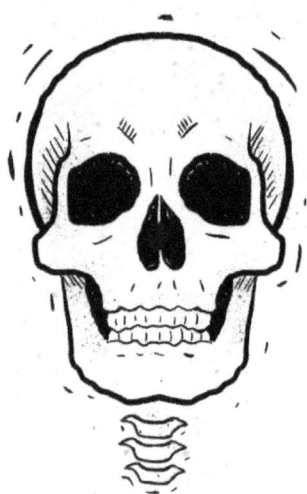

# What in the World are X-Rays?

Written by:
Austin Mardon
Aleefa Devji
Janani Rajendra
Susie Woo
Tara Y.T. Chen
Alicia Au
Sudipta Samadder
Ayah Nour Nehdi
Maria Gonzalez
Eline El-Awad Gonzalez
Megan Ng

Edited by:
Naima Mohamood

**GM**
P R E S S

2021

A Golden Meteorite Press Book.
Printed in Canada.

First Printing: 2021

Typeset and Cover Design by Fariha Khan

Email: aamardon@yahoo.ca
Telephone: 1-(587)-783-0059
Website: www.goldenmeteoritepress.com

Additional copies can be ordered from:
Suite 103 11919 82 Street NW
Edmonton, AB
T5B 2W4
CANADA

ISBN: 978-1-77369-229-6

# Contents

### What Was Medical Imaging Like Before X-Rays?

Medical imaging was invented by Wilhelm Röntgen in 1895 but before the 19th century, he utilized other tools and medical learnings to treat injuries, wounds, or infections based on alternative forms of diagnosis. Today, we use medical imaging and x-rays in various ways and for many reasons in the healthcare system. X-ray technology uses high-energy wavelengths that can pass through objects or the body and allow for the generation of images that show the tissues and structures inside the body. In the present day, we use X-rays to diagnose or visualize respiratory tract conditions, visualize the severity of broken bones or fractures, examine osteoporosis or bone loss severity, diagnose some tumours or arthritis in joints, among other uses. Although x-rays are extremely useful tools to aid in the treatment of patient conditions, they did not always exist. We will explore the history of the diagnosis and treatment of some of these ailments before the invention and use of medical imaging was known. For instance, evidence of bone injuries and bone setting goes back centuries as humans have never been immune to injury and We will explore how bone-setters of the time diagnosed and treated these injuries. As well, the diagnosis of types of pneumonia and respiratory diseases preceded x-ray tomography and stemmed back to Ancient Greek medicine and the Hippocratic Corpus, and humorism.

It is important to understand medicine and health were thought differently in early history. . For instance, humorism played a significant role in medicine, which was explained by body fluids that were the humors (Res Naturalis), and the rules of health hygiene (Res non-Naturalis) (Wiel van der Mark, 2016). This understanding of medicine came from the teachings and insights of Hippocrates and Galen. Described by Hippocrates in 460 to 377 BCE, the humors were paralleled to the four elements of air, earth, fire and water and they each had their properties and likewise the four humours of blood, black bile, yellow bile and phlegm had their own properties. Blood was hot and wet, black bile was cold and dry, yellow bile was hot and dry, and phlegm was cold and moist. Within each person, it was believed that there was a unique combination of these four humours, and they had to be balanced within the individual because they either attracted or repelled each other. Balance of the four humours ultimately meant health, whereas unbalanced humours could lead to illness. In 121-210 CE Galen advanced Hippocrates' theory using anatomical and physiological evidence that showed diseases originated from the inside of the body and resulted from the corruption of the four humours and their balance. During this time, it was believed that injuries, wounds, bone fractures and other infections or diseases could be healed as a result of nourishing actions to balance the humours.

Along with the Res Naturalis or humors, which were natural things occurring from the inside of the body, there were also outside influences on the bodily fluids known as Res non-Naturalis, which were brought about by Galen (Wiel van der Mark, 2016). In brief, these six rules of Res non-Naturalis consisted of:

1. The air we breathe
2. Nutrition, or the balance between eating and drinking.
3. Working and resting and the proper use of exercise.
4. The sleep cycle and being awake.
5. The control of the excretion and retention of bodily fluids.
6. Mastering joy, anger, fear and sadness (those emotions being the passions of the soul).

As we know today, many of our daily activities and the way that we care for ourselves will impact on our health and the way that our body responds to accidents, falls, infectious disease, among other illnesses. Although we are mainly focusing on bone setting and fractures that were diagnosed before the discovery of x-rays, many other illnesses or health conditions such as pneumonia, tuberculosis or arthritis would have been treated based on these humors and outside influences before medical imaging advanced to provide more concrete diagnoses and guidance for treatment.

In particular, when looking at the diagnosis of broken bones before x-ray technology was invented, we must keep in mind that the degree of injury was difficult to measure (Musculoskeletal Key, 2019). This was especially the case with complicated bone injuries such as multiple bone fracture cases. The assessment of bones was done through palpable assessment of limbs or extremities to get a sense of the location of injury. Today, a broken bone often means a visit to the ER but in the 18th century and before there were no x-rays so for most people the individuals who attended to these types of injuries were bone setters who largely relied on common sense in place of medical training. Bone setters would first determine the severity of the injury or if the bone was in fact broken. This process of diagnosis involved examination of the symptoms common to fractures such as a grating noise that is distinguished by movement of the limb that often accompanies fractures, along with the change in form or length of the limb (Bradford, 2014). Basically, their methods involved visual inspection for deformities and if no visual symptoms were present, they would move the limb around while listening for sounds characteristic to bone fractures or rubbing of bone pieces against one another. In most cases where fractures or broken bones were suspected, the bones were pushed back into place and were set in an immobile position for healing to take place.

This information tells us that before X-ray technology diagnosis of these injuries were difficult and required careful observation for how to proceed with patient care. What is not obvious here, is that the wrong diagnosis or decision for how to proceed is made the injuries could be life threatening (Bradford, 2014). In extreme cases where bone injuries were potentially more severe, for instance with a shattered bone or a bone broken into multiple pieces, the bone-setters would have to decide whether the limb should be immediately severed or amputated. The reason for this operation would be to reduce the risk of gangrene or death of the patient, although this is not a skill or practice often done by a bone-setter because of their lack of medical training. As a result, these procedures were often delayed due to hesitation from the bone-setters and as a result were not able to curb the effects of gangrene or possible death.

One of the earliest examples of fracture management for human bones dates back to 3500 BCE in Egypt and was discovered on the Hearst Egyptian expedition in 1903 (Musculoskeletal Key, 2019). These specimens were of splinted extremities using long wooden boards wrapped in linen bandages - unfortunately, this example of splinted bones found during this expedition showed no signs of healing and the victim is thought to have died soon after the injury. The Egyptians were still thought to be skilled at the management of fractures though, as other specimens were found with well-healed bone injuries.

The use of bone splinting with wooden splints and bandaging was used from antiquity to present day and was noted in detail by both Hippocrates and Paracelsus but other techniques of bone splinting were found throughout history as well. In 936-1013 CE an Arab surgeon, Abu'l-Qasim Khalas Ibn Abbas Al-Zahrawi who also went by Albucasis, described a technique of splinting that involved bandaging with 2 layers of fabric, starting at the fracture site and extending up and down the limb (Musculoskeletal Key, 2019). Albucasis went on to discover that alternative materials could be used for splinting such as plaster or mill dust for small women and children along with other materials. Some of his alternative suggestions were plaster mixtures made from various gums, acacia and the root of *mughath* (*Glossostemon bruguieri)* which were pounded together with clay and mixed with water or egg whites. More recently in 1517, Gersdorf illustrated a new method of splinting that involved the binding of wooden splints with ligatures around the splint to hold it tightly together for better support and healing.

Benjamin Gooch went on to copy Gersdorf technique of ligature splinting and used it as a basis for the development of the first brace in 1767 (Musculoskeletal Key, 2019). The brace was designed to help individuals return to their daily activities before the fracture or injury had fully healed. These splints were shaped by Gooch for different anatomical sites and used long pieces of wood, sheets of leather, the ligature design taken from Gersdorf and cannulated bindings to hold them together.

In early 19th century America which was considered the early modern period of medicine, injured limbs were thought to be characteristic of pain and discolouration that may accompany a broken bone (Exploring Illness, 2021). Oftentimes these injuries were confused for contusions or dislocations of the limb because there was no way to observe the underlying issue so whenever it was possible a doctor would be called to diagnose the injury and perform the appropriate procedure or setting. If the injury was in fact a fracture or broken bone, the physician would align the bone fragments as close to the natural state as possible and wrap it in a bandage before splinting it as described above. In some cases, casts made of plaster were used to ensure the limb or bone was well protected from re-injury or further damage.

We heavily rely on medical imaging and x-ray technology for the diagnosis of other illnesses and infections such as pneumonias and respiratory conditions in the present day, but prior to the invention of x-rays we relied on other means of diagnosis and treatment. Going back to Ancient Greek Medicine and Hippocrates there was a high prevalence of lower respiratory tract infection that was confounded with acute illness and mortality for both children and adults (Tsoucalas, 2016). During the time of Hippocrates and his teachings, the infectious manifestations in the lower respiratory tract were diagnosed and categorized through the use of thorough inspection and observation. Some of the manifestations or infections that were seen were bronchiolitis, pleurisy, and pneumonia which are illnesses or infections that would be diagnosed with x-ray technology today. As stated above, Hippocrates and his teachings of medicine and ailments used the theory of the four humours and likewise his description of the pathology of respiratory tract diseases involved these humors.

The use of the humors, Hippocratic opinion and thorough observation allowed for the diagnosis of these illnesses before the invention of medical imaging. Hippocratic opinion explained pleurisy as a pathological condition where the lungs are leaning toward the pleura

(Tsoucalas, 2016). Today we know pleurisy to be inflammation of the pleura which are the two layers of tissue that separate the lung from the chest wall and causes symptoms of severe chest pain or pleuritic pain that is sharp and worsening during breathing (Mayo Clinic, 2021). In hospitals and medical practices today, diagnosis of pleurisy and the underlying cause is determined by imaging studies such as chest x-rays or ultrasounds which are another form of medical imaging. Before this, the Hippocratic followers listened for the characteristic sound of pleurisy otherwise known as pleural friction rub caused by the inflammation of the pleura and subsequent rubbing of the two layers. It was believed that the symptoms which accompanied pleurisy would be fever, pleural pain, and that the mucus would be of a dark colour (Tsoucalas, 2016). Going back to the four humors, alcohol was often blamed for the presentation of pleurisy because wine was thought to cause the deposition of bile and phlegm into the thoracic cavity which would lead to the sticky of the tissues together. As we can see from the teachings of Hippocrates, the diagnosis of illnesses of the lower respiratory tract was done solely by observation of audible signs and observable symptoms but today we largely rely on the use of medical imaging to assist us in making these diagnoses.

In the diagnosis of both bone fractures and lower respiratory tract conditions, the use of auscultation and audible observations played a part in medical diagnosis before the assistance of medical imaging was available. Not long before the invention of x-rays, the stethoscope was invented in 1816 by René Théophile Hyacinthe Laennec who was a renowned French physician and researcher in the field of medicine (Frith, 2021). Initially, he used a rolled-up notebook to assist with auscultation of the pericardium and was able to hear the heartbeat more clearly and distinctly. He first called his invention le Cylindre but later changed the name to stethoscope and coined the method of auscultation "mediate auscultation" which translated to hearing carefully with the aided ear. The stethoscope played a vital role in assisting with the diagnoses of heart and lung conditions by allowing for accurate descriptions of normal function and the comparison to those with abnormal function. To this day, the stethoscope is a vital tool that is used by many medical professionals to diagnose conditions some of which are then confirmed or further examined with the use of medical imaging such as ultrasounds and x-rays.

With the use of his new invention, Laennec was able to further his medical research and provide reliable descriptions of various diseases found in the chest such as bronchiectasis, pneumonia, pleurisy, and emphysema among others (Frith, 2021). As a result of the medical advances that resulted from the use of the stethoscope auscultation became an unequivocally important tool in medical diagnosing as it provided information for what was going on inside the patient. Laennec argued that as a result, techniques of physical examination were superior to the traditional methods of questioning and observation of exterior symptoms. As medicine continued to evolve, and medical imaging along with other diagnostic tools were invented and implemented throughout history we see that Laennec's claims are well supported. Although the stethoscope and Laennec's design faced scrutiny throughout the 19th century, it was improved upon with new materials and designs to combat the criticism and fear and discomfort that deterred patients from the tool.

As we can see from the exploration of common ailments that have existed throughout history such as pneumonia, lower respiratory tract conditions, and bone fractures, we did not always require medical imaging to diagnose and treat conditions. Although medical imaging and specifically x-rays are a key instrument in the medical diagnoses and procedures done in the present day, our ancestors were able to utilize their own methods of observation to treat the diseases, injuries and infections they encountered to the best of their ability. Today, we are fortunate to have precise tools that guide us to diagnosing patients quickly and understanding the details of their conditions to provide the best possible treatment, although as you will learn throughout this anthology there are disadvantages and concerns associated with the use of x-rays and other medical imaging.

## Who Invented or Discovered X-Rays?

The discovery of X-rays has in no doubt been a ground-breaking phenomenon in the diagnosis of many medical conditions. Before X-rays existed, doctors often diagnosed their patients using their best judgement from physical exams . In some cases, doctors had to perform surgery to  see inside a human body (UTMB Health, n.d.; A&E Television Networks, 2009). This approach was not a reliable method of medical diagnosis as it resulted in misdiagnosis or undiagnosed conditions, causing the patient to suffer greatly. From this, one would expect that the discovery of medical imaging was of high priority to scientists and was a deliberate discovery, but that was not the case. Surprisingly, the discovery of X-rays was an accident (Columbia Surgery, n.d.). Although this discovery was unintended, it played a major role in developing and understanding medical  imaging technology that are used today (Columbia Surgery, n.d.). X-rays contributed to the development of magnetic resonance imaging (MRI), computed tomography (CT), ultrasound, echocardiography, and many more appliances (Columbia Surgery, n.d.).

In 1895, Wilhelm Röntgen, a professor of physics in Wurzburg, Bavaria, unintentionally discovered X-rays (Columbia Surgery, n.d.). As a student, Röntgen was expelled from Utrecht Technical School, resulting in his inability  to obtain a diploma (APS physics, n.d.). However, this incident did not prevent him from getting a doctorate nor did it prevent him from conducting experiments about light phenomena and other emissions at the University of Würzburg (APS physics, n.d.). Röntgen had a great interest in cathode rays and his experimentation with cathode rays is what ultimately led him to his unintentional discovery of X-rays (APS physics, n.d.).

On November 8, 1895, Röntgen made a notable observation while experimenting with cathode rays and a glass tube (Columbia Surgery, n.d.). When the tube was covered with heavy black paper, green incandescent light was transmitted onto a fluorescent screen (Columbia Surgery, n.d.). After examination, he realized that the tube was releasing an invisible ray that penetrated the black paper surrounding the tube (APS physics, n.d.). At first, he thought that these invisible rays were unrelated to light as they did not demonstrate any characteristics of light, such as reflection or refraction (Encyclopædia Britannica, inc., 2021). As such, due to the ambiguous nature of these invisible rays, he decided to call them "X" rays, as "X" is meant to denote "unknown" (Columbia Surgery, n.d.). Thus, X-rays were also referred to as X-radiation and Röntgen radiation (Encyclopædia Britannica, inc., 2021).

Röntgen spent seven weeks alone, figuring out his discovery, all while keeping his work a secret from everyone (University of Washington, 2018). Then, he conducted further experiments which led him to note that the X-rays were able to pass through most substances except solid objects, which is what caused shadows (Columbia Surgery, n.d.). Shortly after, he found that X-rays could pass through human soft tissue, while bones and metals remained visible (Columbia Surgery, n.d.; APS physics, n.d.). The first X-ray images were taken of his wife's hand and metal objects (Encyclopædia Britannica, inc., 2021). The discovery of X-rays was revolutionary in the medical field. It soon became popular worldwide as it allowed for not only medical imaging of the human body, but also for treatment of diseases (APS physics, n.d.).

Wilhelm Röntgen was awarded the Nobel Prize for Physics for his discovery of X-rays (Encyclopædia Britannica, inc., 2021). He remained quite humble and altruistic throughout his entire life as he did not want honour or financial profits from his research and discovery of X-rays (APS physics, n.d.). He rejected honourable titles and donated his Nobel Prize money to his university (APS physics, n.d.). Further, Röntgen did not file any patents for X-rays as he believed that the public should be able to benefit from his work at no monetary cost (APS physics, n.d.). This allowed many other researchers to experiment with X-rays which ultimately led to the creation and advancement of science and technology.

As time went on many other researchers such as Thomas Edison, William D Coolidge, Henri Becquerel, Marie Curie, Pierre Curie, Otto Walkhoff and Fritz Giesel experimented with X-rays and made significant contributions to the advancement of X-rays. These researchers set the basis for radiation imaging, and without these greatly talented researchers, many medical diagnostic imaging methods and treatments would not exist today.

When X-rays were discovered in 1895, Thomas Edison was a well-reputed individual in society and began to experiment and explore X-ray radiation (Michigan State University, n.d.). At that time, X-ray tubes were not very accessible; therefore, Edison created his X-ray tubes since he had prior experience manufacturing incandescent lights similar to electric light bulbs (Michigan State University, n.d.). Edison's experimentation and research on X-rays were dedicated to improving the barium platinocyanide fluorescent screens that were used to view X-ray images discovered by Röntgen (Michigan State University, n.d.). After conducting several experiments, Edison found calcium tungstate to be a more effective material to view X-ray images than barium platinocyanide (Michigan State University, n.d.). Edison added calcium tungstate into a device called the Vitascope (fluoroscope). This invention by Edison quickly became a popular tool used by doctors for viewing X-ray images (Michigan State University, n.d.). Clarence Dally assisted Edison throughout all his X-ray research. Unfortunately, due to high radiation exposure, Dally developed a skin disorder that led to carcinoma, ultimately resulting in death (Michigan State University, n.d.). Dally was the first person known to have passed away from radiation in the United States in 1904 (Michigan State University, n.d.). For this reason, Edison decided to end his X-ray research and investigation as he realized that X-rays are poisonous (Michigan State University, n.d.).

William D Coolidge made a vast contribution to radiology when he discovered the X-ray tube, which is now the basis of every X-ray machine (Bell, n.d.). Prior to Coolidge's contribution to the X-ray tube, all X-ray tubes were made from the Crookes gas tube design, which was known to have many limitations (Bell, n.d.). A substantial flaw of the Crookes gas tube occurred when electrons were emitted and struck platinum to produce X-rays (Bell, n.d.). The collision of electrons with the platinum caused vaporization and condensation of platinum on the glass leading to diminished efficiency of the Crookes gas tube (Bell, n.d.). The tube designed by Coolidge used tungsten and varied the amount of current in the tube, which affected the number of electrons and the X-ray radiation emitted (Bell, n.d.). The ability to change the voltage of the tube allowed for control over penetration of X-rays (Bell, n.d.). Coolidge's tube was stable, accurate, consistent, versatile, durable, powerful, and overall, much safer than the original tube, making it much more desirable than the original Crookes gas tube.

As cameras and X-ray machines developed, multiple different objects could be put on a single sheet of paper (Frankel, 1996). This development reduced the cost of X-ray imaging significantly and reduced the time taken to take each X-ray image (Frankel, 1996). Further, more efficient x-ray imaging methods were being developed; researchers were also looking to find ways to reduce the amount of radiation exposure per X-ray image (Frankel, 1996). Reducing the radiation exposure time not only sped up the X-ray imaging process, but also reduced the amount of radiation each patient was exposed to (Frankel, 1996). The solution to this came from the development of intensifying screens (Frankel, 1996). Therefore, intensifying screens are placed on each side of the X-ray film, and when they are hit with X-rays during X-ray imaging, the radiation is transformed into light energy (Frankel, 1996). Since the X-ray film is sensitive to light energy, the emitted light energy will ultimately produce the final X-ray image (Frankel, 1996). Intensifying screens lowers the amount of radiation emitted per X-ray image compared to when no intensifying screens are used (Frankel, 1996). As such, intensifying screens successfully reduced the toxicity caused by X-ray radiation to patients (Frankel, 1996).

In the 19th century, tuberculosis caused many deaths (Frankel, 1996). X-ray imaging was used to diagnose patients with tuberculosis (Frankel, 1996). The use of X-rays allowed for earlier detection of tuberculosis, which allowed patients to enter isolation early on, ultimately reducing the spread of this disease (Frankel, 1996). Since X-rays were both low in cost and efficient, X-rays played a predominant role in controlling the spread of tuberculosis in the 19th century (Frankel, 1996). In like manner, X-ray imaging also contributed to the prevention of tuberculosis and the creation of antimicrobial therapy, which was used to treat tuberculosis infection (Frankel, 1996). Even after controlling the mass tuberculosis infections, the use of X-ray imaging for diagnostic purposes continued to increase rapidly (Frankel, 1996).

A few months after discovering X-rays, Henri Becquerel discovered radioactivity (Tretkoff, n.d.). Becquerel studied fluorescence and phosphorescence and had a particular interest in learning uranium (Tretkoff, n.d.). After hearing about Röntgen's discovery of X-rays in 1896, Becquerel started investigating the relationship between phosphorescence and X-rays (Tretkoff, n.d.). He hypothesized that phosphorescent uranium salts could absorb sunlight and radiate it back as X-rays (Tretkoff, n.d.). At first, Becquerel thought that sunlight was needed to create X-rays. However, as he conducted an experiment without the presence of the sun t, he was stunned to see that the X-ray image was remarkably clear (Tretkoff, n.d.). After this observation, he conducted several more experiments and confirmed that uranium salts had the ability to emit radiation on their own, in the absence of sunlight (Tretkoff, n.d.). Just like Röntgen's unintended discovery of X-rays, Becquerel's discovery of uranium emitting radiation in the absence of sunlight was accidental as well (Tretkoff, n.d.). About 40 years prior to Becquerel's discovery, Abel Niepce de Saint Victor who was a photographer, investigated many chemicals including compounds that contained uranium (Tretkoff, n.d.). Similar to Becquerel's experiments, Niepce examined uranium and other compounds both in the presence and absence of sunlight and thought that he discovered a new invisible radiation (Tretkoff, n.d.). Niepce declared his findings to the French Academy of Science, but no one conducted further research on the topic until Becquerel repeated the same experiment in 1896 (Tretkoff, n.d.). During Becquerel's experimentation he also found that non-phosphorescent uranium compounds were able to emit radiation as well (Tretkoff, n.d.). Initially, Becquerel thought that his rays were similar to X-rays; however, the rays emitted by uranium had a different property which allowed uranium rays to be deflected by electric or magnetic fields, making them different from X-rays (Tretkoff, n.d.).

In 1898 Marie Curie and Pierre Curie began studying uranium rays (Tretkoff, n.d.). They were able to measure the intensity of radioactivity and they also discovered new radioactive elements including polonium, thorium, and radium (Tretkoff, n.d.). Marie Curie defined the word radioactivity, formalizing it to be radiation evoked by atomic decay (A&E Television Networks, 2010). Marie Curie won two Nobel Prizes for her achievements. Soon she began creating radioactive substances that could be used for medical purposes.

During World War I, X-ray machines were only found in city hospitals, which made them inaccessible to injured troops on the battlefield. Marie Curie invented the first transportable vehicle containing an X-ray machine and a photogenic darkroom equipment. This portable X-ray machine was able to be driven around the battlefield to help doctors diagnose and treat wounded individuals more conveniently (Jorgensen, 2017). Marie Curie overcame several obstacles such as the need for a power source for the X-ray machine, the lack of funding for her invention and the need for trained X-ray operators (Jorgensen, 2017). She courageously handled the challenges that came her way, and the X-ray vehicles were used during the first world war. These vehicles were known as the mobile little Curies (Jorgensen, 2017). Even though the X-rays vehicles helped significantly on the battlefield, there were some unwanted casualties due to the lack of knowledge about X-ray radiation at that time. As a result, many X-ray workers, including Marie Curie herself, suffered adverse effects from being overexposed to X-ray radiation (Jorgensen, 2017). On the other hand, over one million injured soldiers used Marie Curie's X-ray vehicles during World War I. Marie Curie made a substantial impact in medical and science research. However, she died in 1934 from leukemia caused by her many decades of exposure to radioactive materials (A&E Television Networks, 2010).

X-rays were first used in 1897 during the Balkan War in order to find bullets and broken bones inside humans (A&E Television Networks, 2009). Soon enough, X-rays became a very popular method to diagnose patients due to the tremendous benefits it provided (A&E

Television Networks, 2009). However, the adverse effects caused by radiation of X-rays were unknown at this time (A&E Television Networks, 2009). A couple years after working with X-rays, Röntgen observed hair loss and skin toxicity (Khare et al., 2014). Röntgen unknowingly protected himself from radiation by holding a lead plate between his fingers whilst working with the X-ray radiation (Khare et al., 2014). Unfortunately, many other researchers and scientists suffered tremendous consequences due to the exposure of radiation while experimenting with X-rays (Khare et al., 2014). Years later, cases of burns, skin damage, and skin cancer caused by exposure to X-ray radiation were being reported (A&E Television Networks, 2009).

The invention of the first dental radiographs required the inventors to be exposed to radiation for a significant amount of time–25 minutes (Sansare, 2011). Walkhoff described his experience as being torturous (Sansare, 2011). After further experimentation, Walkhoff succeeded in creating oral X-ray images (Sansare, 2011). Accordingly, Otto Walkhoff and Fritz Giesel created the first dental roentgenological laboratory (Sansare, 2011). They imaged heads and jaws in their laboratory for doctors and due to this, they were greatly exposed to the radiation given off from the X-rays (Sansare, 2011). Sadly, in 1927, Fritz Giesel passed away due to metastatic carcinoma due to exposure of heavy radiation on his hands (Sansare, 2011).

Everyone experiences radiation on a daily basis (Newman, 2018). Radioactive material occurs naturally in air, soil, water, rocks and vegetation (Newman, 2018). The radiation given off from these sources are very little, and thus are not harmful (Newman, 2018). People working occupations that involve exposure to cosmic rays at altitudes are at a greater risk of developing cancer as they are exposed to high doses of radiation (Newman, 2018). Unfortunately, many researchers that participated in the creation and development of X-rays were poisoned by the radiation to some extent, which ultimately resulted in their death. However, the contributions made by these researchers are extraordinary and they helped pave the path to where we are today in terms of science and technology.

## How Did the Discovery of X-Rays Impact Medical Research and Imaging?

The novel discovery of x-rays in 1895 gained momentum in further studies and research, where within the first year of it's forthcoming, resulted in widespread applications in the fields of surgery and orthopedics. Initially, with the radiograph (i.e. the x-ray) demonstration of the hand of Röntgen's wife, Bertha, it was evident that x-rays had the potential to image dense objects invisible at surface level and reflect them as opaque images on film. This included bones and metallic items, such as bullets and other types of shrapnel (Murphy, 1990). In the year of 1896 alone, there were a total of 49 books and brochures, as well as 1044 scientific essays that addressed the potential scientific applications of x-rays, with multiple studies researching diagnostic medicine (Banerjee et al., 2013). In the early stages of radiography, the imaging technique by use of x-rays, was mainly directed towards studying the human skeleton in specific areas of the extremities. Usually the hands and feets were the focus of research as they didn't have multiple layers of tissue covering the bones. The first medically indicated radiograph was conducted by Gustav Kaiser and Eduard Haschek, where they were able to pinpoint trauma to the middle phalanx of their patient.

Although radiography was immature in its beginnings, the technology was sufficient to detect foreign bodies with a high degree of effectiveness. For instance, on January 19, 1896, an Austrian physician, Dr. Pfaundler, was able to locate a needle in the hand of a 12-year old girl using x-rays, which could not have been done otherwise with the naked eye. In addition, fracture detection was also an application of radiography that could aid physicians in the analysis of bone diseases (Banerjee et al., 2013). In February of 1896, F. Konig, a Berlin physician, correctly diagnosed a patient of having sarcoma in the tibia by using x-ray imaging (Strickland & Stranges, 1991). However, the cost of this pathological find was only possible with the amputation of the limb for a proper x-ray photograph (Banerjee et al., 2013). Another forerunner of radiography was Charles Thurstand Holland, who compiled and collected hundreds of photographs regarding the musculoskeletal system. Majority of his findings were related to fractures and dislocations, which allowed him to gain further understanding of various congenital deformities. Furthermore, his contribution elucidated the growth and development of the hands in children, which contrasted research that solely focused on acute diagnoses. Similar to Konig, Holland was also exceptionally involved in examining pathologic conditions which included rheumatoid arthritis, hypertrophic osteoarthropathy, tuberculous arthritis, enchondroma, osteochondroma, and rickets (Murphy, 1990).

On a macroscopic scale, there were a few individuals who were able to successfully capture an x-ray image of the entire human skeleton. The first person to accomplish this feat was none other than Röntgen's assistant, Ludwig Zehnder. In the August summer of 1896, Zehnder photographed body parts of several people with exposures ranging from 5 to 15 minutes, which were then organized to create a close mimicry of the human skeleton in its entirety (Farmelo, 1995). Another notable mention of what was closely considered to be a whole body image, was attributed to J. Hall Edwards of Queen's Hospital in Birmingham, England, who was able to produce skeletal photographs of the backbone, vertebrae and spinal cord (Strickland & Stranges, 1991). Nonetheless, the limitations of full-body imaging was due to technological constraints and imaging techniques. However, in 1897, William James Morton, one of the most ambitious pioneers of radiography in America, was able to create the first successful single-film of the whole body skeleton. The subject of interest was a 30-year old woman who was 5 feet 4 inches in height and weighed 120 pounds. Through the utilization of several preliminary induction coils and tubes by Edison and Crookes, exposure to the radiation took 30 minutes for completion (although the process included frequent stoppages to permit tube cooling). This contrasts later developments in the 1930s, whereby a man named Arthur Wolfram Fuchs significantly refined the whole-body imaging process by using selective filtration and screens. His method required a higher voltage and current which resulted in less exposure time to only 1-2 seconds. The potential of x-ray diagnostics was profound in its instigation of advancements into new streams of medical imaging, which included nuclear medicine, magnetic resonance imaging (MRI) and computed tomography (CT) scans (Murphy, 1990).

The application of x-rays was momentous in times of conflict and war, as to prevent further infections that festered from gunshot wounds and bacteria that were pervasive on the battlefield. In particular, Madame Curie's contribution in medical radiology during the first world war, undoubtedly laid the foundation in the adoption of x-rays for diagnostic and potentially, therapeutic purposes. In 1914, she was able to get funds from donors and manufacturers to assemble 20 vehicles with x-ray equipment to be used in the war. The electricity necessary to power the x-rays were gathered from the engines of the trucks. This was a revolutionary idea during wartimes as in previous battles, it was impractical to bring radiation technology to the front lines due to a lack of a consistent source of electricity (Martin, 2018). With her idea, responders were able to provide immediate surgery and save lives with precision and timing (Halliday, 2017). Furthermore, Madam Curie established a total of 200 stationary x-ray facilities in the rear, in which they also had a stable energy source. Although these facilities were not on par to modern technology in today's world, they were sufficient in identifying bone fractures and splinters, as well as shrapnel embedded in the bodies of soldiers. The x-rays that depicted these sorts of injuries amounted to over 2000 photographs (Martin, 2018). The sheer volume of x-ray usage, and the improved recovery rate of soldiers demonstrated the importance of an accurate diagnosis to systematically saving lives. After the first world war, it was evident that proper medical imaging can vastly improve the chances of a patient's survival, and had an influence in the adoption of radiology as a speciality in the field of medicine.

One of the greatest implications of x-rays was the ability to detect tuberculosis during the second world war. Based on experience from World War 1, tuberculosis was an incurable, respiratory disease that resulted in some misdiagnoses amongst drafted soldiers. On top of the poor, unhygienic condition of the battlefield, soldiers were more susceptible to widespread contraction of the disease. Since there was no cure, patients who were suffering from tuberculosis were withdrawn from the war and were placed in sanatoriums for isolation until they recover by themselves or die in their beds. The United States themselves paid approximately $960 million towards care for World War 1 veterans who contracted tuberculosis during their military service. Hence, it was impervious that proper screening of soldiers was necessary to prevent widespread contagion of tuberculosis. Radiographic technology was the key to assessing lung opacity, form, and size, which contributed to the overall diagnosis. Chest x-rays that were rarely conducted in the first world war, were now a requirement for prospective soldiers in both the army and navy. Even though there was a shortage of radiologists and equipment available for screening, this process allowed physicians to correctly diagnose those who had no or minimal outward symptoms of the disease. Approximately 1% of individuals from the armed forces were warranted a rejection due to being diagnosed as positive for tuberculosis. However, the rate of rejection was unexpectedly high, and raised suspicion amongst the authorities. This could be attributed to the new, arbitrary standards of assessment for tuberculosis, which relied on the presence or absence, number and size of calcified parenchymal nodules and intrathoracic lymph nodes (Haygood, 1994). Later studies have shown that chest x-rays may not be the best choice for complete diagnosis of the disease. In fact, skin tests were the most useful mass screening method where positive cultures for *M tuberculosis* was the only positive proof of the active form of the disease (Husen, Fulkerson, Vecchio, Zack, & Stein, 1971). However, just as chest x-rays have some limitations, skin tests are not absolutely foolproof either. It is important to recognize both methods as imperative to tuberculosis testing and should be used according to well-researched, national standards in modern medicine.

Although the discovery of x-rays in allowing visibility of the human bones and foreign objects was significant in and of itself, there were still many internal organs and soft tissues that needed to be explored in medical imaging. British pioneer John Macintyre succeeded in taking an x-ray of the thorax to analyze the heart and lungs. Due to the poor performance of the x-ray unit, the radiographs were blurry and unclear. They could not provide clarity on the form, size, and position of internal organs, and surrounding fluids. The translucence of the lungs and heart, as well as shadows caused by pleural fluid inhibited proper imaging of the chest cavity (Banerjee et al., 2013). Contrast media are substances that can be injected into the body to make fluids and internal organs visible in x-ray photographs. Hence, they are crucial in investigative radiology for mapping accurate depictions of less dense, internal structures. One of the first contrast agents that was accepted for clinical use was iodine-containing compounds, as it was deemed to be sufficiently safe for human ingestion. The development and synthesis of the ideal contrast agent continued in the 1920s, but there were difficulties with reducing the toxicity, osmolality and viscosity of the agents to minimize detrimental side effects and feelings of pain during injection. In studies done by Lasser et al., the mechanisms by which these adverse effects develop, lies in the hydrophobic bonds of the contrast agents to biomacromolecules. The level of viscosity of the radio contrast agent makes it very painful in intra-arterial injection, and so scientists have tried various combinations of both ionic and non-ionic contrast agents with the least amount of side effects (Quader, Sawmiller, & Sumpio, 2000). One notable application of opaque substances occurred in 1919, where physician Carlos Heuser found that water-soluble iodine compounds made the veins and arteries opaque to x-rays. This was further used in urology to study the urinary tract, as the blood (which contained these agents) filters through the kidneys. More specifically, the compound tetraiodophenolphthalein is used to diagnose gallbladder issues and was later known as the Graham-Cole test. Similar contrast media allowed the diagnosis of tumors or lesions in the spinal column by using a hydrophobic, iodine-based substance called lipiodol. By injecting this agent into the subarachnoid space of the spinal column, physicians were able to trace the migration of the substance in the spinal fluid. Any blockages that were shown through the radiographs indicated the presence of a tumor or lesion. Lipiodol is a good contrast agent that provided well-documented x-ray photographs, however it had a slow absorption rate and irritated the tissues within the subarachnoid space. Moreover, this applies to most halogen-based compounds as they were well known to also cause irritations and problems in the gastrointestinal tract (Strickland & Stranges, 1991).

In the 1890s, physicians commonly prescribed bismuth-based medication to treating gastric ulcers. However one physician named Pfahler realized the potential for bismuth as a contrast agent, similar to halogens. However, bismuth was a better alternative as they were unlikely to cause chemical reactions with the gastric fluids, and thus did not result in abdominal discomfort. Pharmaceuticals began to prepare bismuth subnitrate, which was a white, water insoluble, non-toxic compound that became a standard choice for digestion research. In fact, Harvard Medical School professor, Walter B. Cannon was able to monitor the actual movement of the digestive system by orally administering bismuth subnitrate in both animal and human subjects. Although there were some reported adverse side effects in his patients, he discovered another compound, barium sulfate, which produced the same results and was physiologically safer. Barium sulfate instantly replaced bismuth subnitrate in 1910, whereby it's low cost production and chemical safety made it a well-known substitute. Cannon's discovery in the movement of the digestive system gave rise to cineradiography, whereby rapid succession of x-ray photographs roughly resembled the actual motion of the gastrointestinal tract (Strickland & Stranges, 1991).

Another key organ of interest to physicians at the time, was the brain. Normally, it was difficult to distinguish on film, due to its low opacity like the other internal organs, however a professor of neurology at the University of Lisbon, Antonio Egas Moniz, was able to apply the use of contrast mediums to tackle this issue. He discovered that injecting sodium iodide into the inner carotid artery allowed the depiction of x-ray photographs of the brain's arterial network. This allowed him to locate deformations in the brain caused by tumors (Strickland & Stranges, 1991).

Hence, throughout the development and research x-ray techniques, physicians were able to view internal organs and tissues in addition to musculoskeletal mass. This enabled doctors to gain further insight on diseases in various fields within medicine, such as cardiology, urology, and gastroenterology. However, the speciality of obstetrics in particular had garnered interest by the public in the 1920s. At the time, ultrasonography did not become a routine clinical practice in America, and thus, physicians relied on radiography instead. This allowed them to confirm pregnancies by locating the fetal osseous structures, assess fetal position, estimate gestational age, and check for fetal bone anomalies, such as achondroplasia - a rare bone growth disorder that prevents the transition of cartilage to bone (Genetic and Rare Diseases Information Center, 2017). Not only does radiography allow physicians to monitor the fetus' bone health, but also the mother. To ensure no problems occur during labor, the maternal pelvis is examined for preparation during childbirth. Unfortunately, x-rays were still under it's infant stages whereby researchers have not yet fully realized the harmful repercussion of extensive exposure to radiation. It was only until 20 years later when medical professionals realized the potential harm of this form of radiation on the fetus' health (Benson & Doubilet, 2014). There was an increased rate of serious fetal abnormalities, such as microcephaly - a condition by which the baby's head is smaller than expected - mental retardation, microphthalmus - a condition in which a baby is born with underdeveloped eyes - cataracts, and delays in skeletal and sexual maturation. This was seen in several case reports that examined pregnant women at varying stages of pregnancy. Furthermore, the Hiroshima and Nagasaki atomic bomb explosion provided additional data in which children of mothers who were within 1200 meters of the bomb hypocenter in Hiroshima had microcephaly and mental retardation. In contrast, children of mothers who were farther away from the explosion did not have these abnormalities (Dekaban, 1968). These disorders were commonly seen in neonates of women who were exposed to radiation after conception compared to those of women irradiated prior to conception. Hence, it is recommended to limit the exposure of x-rays during pregnancy and should only be used for diagnostic purposes. Furthermore, x-rays were not recommended for pregnant women who were in the critical period of fetal development, which was around 4 to 8 weeks of gestation. Even though warnings were administered, maternal pelvimetry and fetal cephalometry was necessary to prevent complications during labor. The use of x-rays were necessary to gauge the size of the birth canal for appropriate passage during labor, determining low-lying placenta location, and amniography. In addition, some physicians employed the use of contrast agents in the amniotic cavity to check for fetal swallowing, and molar pregnancies (Benson & Doubilet, 2014).

The field of radiography was seen in a positive light in most parts of the literature, and hence, the negative impact of x-rays over prolonged periods of time were not explicitly studied. For instance, x-rays were not only used for taking pictures of the human body, but they were also used as therapeutic treatments for certain diseases. One example can be seen in the usage of radiotherapy as treatment for gas gangrene in the early 20th century. The bacterium, *Clostridium welchii* (also known as *Clostridium perfringens*) was the cause of a progressive and often fatal disease that was found in war wounds or in people who lived in unhygienic conditions. The first formal report of the therapeutic uses of x-rays was noted by Dr. James F. Kelly. As a radiologist, he was fascinated by his discovery of some of his patients recovering from gas gangrene after receiving low doses of x-rays. One dose was approximately 75 rads/day and on average, patients received 2-3 doses depending on the severity of the wound. At the time, amputation was the only method for saving a patient with gas gangrene, however, even this method was not effective for those with the acute toxic phase of the disease. Radiotherapy on the other hand, was shown to have some alleviating effects. This was a significant discovery because prior to World War 1, mortality was approximately 50% of those who contracted the disease. Although radiotherapy was a widely accepted method of treatment in the 1930s to 1940s, proving the direct impact of x-rays on gas gangrene was ambiguous, as most cases were infrequent, highly individualistic, and had minimal statistical influence. Even the American military stated that they found no evidence of x-rays on the therapeutic treatment of gas gangrene, and was not used by the British forces either. After the second world war, treatments for gas gangrene included serum antitoxin therapy and chemotherapy, but not radiotherapy. Literature studies that refuted Kelly's claim have shown that there are debilitative consequences on health on prolonged radiation exposure, and can even lead to further tissue damage (Calabrese & Dhawan, 2012).

In conclusion, there are several advantages of the development of x-rays in medical imaging prospects. It allowed physicians to visualize the human skeleton and foreign objects which provided great insight in surgical procedures. Moreover, the innovative usage of contrast media further advanced radiographic imaging to include internal organs and tissues that were once translucent on film. Although there have been many unfortunate cases of patients suffering from repeated exposure to the radiation across all fields in medicine, the findings were hardly insignificant in pioneering towards better implementation of x-rays in the future. The rapid acceptance of this field played a major role in creating a niche for radiology and specialized practice of diagnosing health issues ranging from orthopedics to neurology. The impact of x-rays on medical research and imaging provided a new area in clinical practice that establishes a solid foundation for the development of revolutionary surgical procedures and disease testing that were once not possible prior to its inception.

## What are X-Rays and Why Are They Worth Investigating and Developing?

### Introduction

X-rays allow for non-invasive imaging of the body using electro-magnetic radiation. An X-ray is a high-frequency energy wave that is either absorbed or transversed through internal structures of the body. The X-ray pattern produced is transferred to a detector for evaluation of the X-ray image on a computer (Nakashima & Duong, 2021). The colour contrast of the X-ray image is dependent on the tissue density of the internal structures. Dense structures, such as bone or metal implants, tend to absorb the most radiation and appear white on an X-ray image. Whereas, soft tissue, such as muscles or organs, allow X-rays to transverse through them and appear gray on an X-ray image. Air is depicted as black on an X-ray image because it absorbs the least amount of X-ray (Chen et al., 2013). X-rays are commonly used in medical imaging technologies to diagnose and manage a variety of health conditions. These technologies have shown an increasing global demand and continue to be a primary tool for managing rare diseases, analyzing head and neck conditions and performing chest X-rays (Ngoya et al., 2016). However, X-ray imaging techniques have limitations and may be improved through the incorporation of artificial intelligence (Hashmi et al., 2020).

### Types of imaging technologies that use X-rays

X-rays are used in a variety of medical imaging technologies, each with their own purpose. The three categories of X-ray imaging are radiography, fluorescence and computed tomography (Tafti & Byerly, 2020).

X-ray radiography produces a static image and is the most popular method of X-ray imaging used in clinical settings. Radiography
uses the least amount of radiation compared to the other two X-ray
imaging methods, which reduces the amount of radiation-associated health risks to the patient. Radiography is also low-cost, able to
provide portable examinations and universally available (Tafti &
Byerly, 2020). Radiography is commonly used for the diagnosis of
chest-related conditions such as pneumonia or pulmonary edema.
Chest and abdominal radiographs are useful to aid in evaluating
the positioning of support devices such as surgical drains (Tafti &
Byerly, 2020). Notably, radiographs are important for evaluating the
musculoskeletal system and are commonly used to evaluate for arthritis, bone fractures and bone tumours. Mammography is a subset
of X-ray radiography used to assess the internal structures of breasts.
However, the visibility of internal structures from X-ray radiography
may be limited in some situations and result in inaccurate diagnoses
(Tafti & Byerly, 2020).

Fluoroscopy uses X-rays to conduct real-time imaging of internal
structures through the use of contrast agents (Tafti & Byerly, 2020).
A contrast agent can absorb X-rays to increase the contrast of an
X-ray image and improve the clarity of the internal structures of the
body (Mahajan & Singh, 2019). An iodinated contrast agent is delivered intravenously to the patient's blood vessels to assess blood
flow in the vascular system through a process called fluoroscopic
angiography. A barium-based contrast system is delivered orally to
the patient to evaluate the anatomy and cause of diseases related to
the gastrointestinal tract (Lusic & Grinstaff, 2013). Moreover, X-ray
fluoroscopy can be used to assess the condition of the biliary system,
which includes the bile ducts, liver and gallbladder. It can also aid in
image-guided diagnostic and therapeutic joint injections of the musculoskeletal system (Tafti & Byerly, 2020). Although the production
of real-time images are helpful for aiding clinicians in the diagnosis and treatment of certain conditions, X-ray fluoroscopy produces
relatively high amounts of radiation and contrast agents may induce
mild to severe-side effects. Mild side effects of using contrast agents
occur with an incidence of less than 3% and include nausea, skin

rashes and dizziness. Severe side effects of using contrast agents occur with a lower incidence of less than 0.04% and include facial swelling, hypertension, coronary artery spasm, cardiac arrhythmias and pulmonary edema (Mahajan & Singh, 2019).

Computer tomography (CT) produces several X-ray images of the target tissue as a detector coupled with the X-ray source rotates around the patient's body. A computer reconstructs all the X-ray images into different planes to visualize the target tissue in a three-dimensional space, which allows clinicians to view the internal structure at any angle (Tafti & Byerly, 2020). CTs are able to produce non-destructive three-dimensional images of a variety of internal structures including the liver, lungs, structures of the cardiovascular system, bones, gastrointestinal tract and tumorous tissue (Lusic & Grinstaff, 2013). This process only takes a few minutes to complete and has a higher contrast sensitivity compared to radiography, which is the reason why CT is the preferred imaging technology to evaluate soft tissues (Nakashima & Duong, 2021). In addition, advancements in recent CT developments have resulted in CT technologies becoming less expensive, less time-consuming and more readily available compared to non-X-ray dependent technologies such as magnetic resonance imaging and positron emission tomography. It is projected that further advancements to CT development may continue to improve the accessibility and quality of X-ray imaging results from CT (Lusic & Grinstaff, 2013). Similar to fluoroscopy, the ability of CT to produce images with increased visibility and differentiation amongst internal structures is due to the use of contrast agents (Mahajan & Singh, 2019). It is estimated that the usage of contrast agents can improve image clarity by a factor of two (Lusic & Grinstaff, 2013). However, similar to fluoroscopy, the use of high doses of contrast agents may be harmful to the patient and result in various mild to severe-side effects. Compared to radiography, there is also an increased radiation risk due to the requirement of producing several X-ray images of the target tissue (Lusic & Grinstaff, 2013).

Overall, X-ray imaging techniques can be categorized into three groups: radiography, fluorescence and computed tomography. They can be used to diagnose, treat and evaluate a large variety of internal systems and organs, which have contributed to advancements in medicine and provided affordable and efficient healthcare to patients globally (Tafti & Byerly, 2020).

## Increasing global demand for improved X-ray imaging technologies

In the past half-century, advancements in the field of imaging technology have resulted in improvements in X-ray radiography, fluoroscopy and CTs. The improvements to this field have increased the value of radiology to patient care and enhanced the productivity of healthcare systems globally. X-ray imaging technologies are considered mandatory for the delivery of primary care because of its long history of being an essential method of diagnosing conditions, assessing disease progression, contributing to preventative health care programs and evaluating treatment responses (Ngoya et al., 2016).

For the past several decades, the usage of X-ray imaging technologies has shown an increase in demand. Between 1988 and 2008, the use of imaging technologies increased by two-fold on a global scale. In 2016, imaging contributed to 10% of the total per capita healthcare expenditure (Ngoya et al., 2016). The increasing demand for X-ray imaging technologies is a consequence of an increased human life-expectancy, global population growth and the rise of chronic diseases. It is expected that X-ray imaging technologies will continue to be in demand in the future, which may put a strain on these expensive and labour-intensive services (Ngoya et al., 2016).

Currently, the demand for X-ray imaging services exceeds the capacity of healthcare systems, particularly in low-income and middle-income countries. For instance, Tanzania has only 5.7 radiography units per million people compared to the recommended 20 radiography units per million people, which indicates a need for more cost-effective imaging technologies (Ngoya et al., 2016). Although X-ray imaging technologies are useful in lower income countries for the management of conditions such as pneumonia, HIV, and tuberculosis, there is a limitation in the quality of X-ray images produced. Countries with fewer resources have cheaper X-ray imaging techniques, which produce lower quality images. Subsequently, the sub-standard

X-ray images delay treatment delivery and reduce the productivity of healthcare services (Zennaro et al., 2013). Typical X-ray techniques use screen-films to produce images, which is expensive and requires additional post-production techniques to improve the image quality. The extra post-imaging machinery and requirement of specialized staff for film processing contributes to the high cost of using X-ray imaging technologies, which is not financially supported in low-income and middle-income countries (Zennaro et al., 2013). It is suggested that a digitized version of X-ray imaging can eliminate the costs of film processing, film development, and specialized personnel. Digital technology is easier to use, more time-efficient, financially sustainable and produces higher quality images (Zennaro et al., 2013). Global inequities for access to high-quality X-ray imaging technologies should be addressed in order to decrease the child mortality rate, increase the quality of life and provide more effective treatments for people living in lower income countries (Ngoya et al., 2016).

<u>The use of X-ray imaging to diagnose and treat health conditions</u>

While X-rays are useful for diagnosing and treating several conditions of the human body, X-ray imaging technologies are particularly important for the management of rare diseases. A rare disease is a condition that affects fewer than 2000 people worldwide and 36% of rare diseases affect the musculoskeletal system. For example, Duchenne muscular dystrophy is a rare disease that affects the muscles, Charcot-Marie-Tooth disease affects the nerves and osteogenesis imperfecta affects the bones (Rossi et al., 2019). There are between 5000 and 8000 rare diseases that are chronic conditions and lead to progressive deterioration of the body. These rare conditions are often under-researched, difficult to diagnose and commonly misunderstood, which leads to inefficient disease treatment and accelerated worsening of the patient's health (Rossi et al., 2019). Luckily, X-ray imaging is a crucial tool that has improved the prognosis of rare diseases for patients in the past. The use of X-ray imaging technologies, such as CTs, can provide a timely diagnosis and appropriate monitoring of treatments. For rare diseases, 53% can be diagnosed with CT and 64% can be diagnosed with X-ray radiography. These X-ray imaging techniques have been one of the primary contributors to the rapid improvements in the field of rare disease research in the past few decades (Rossi et al., 2019). However, despite the significant advancements of rare disease management, 95% of conditions remain without proper treatments and patients experience a 5-30 year delay in diagnosis. It is projected that improvements in X-ray imaging technologies can provide earlier diagnoses, reduce the number of untreatable rare diseases and relieve the strain placed on healthcare systems (Rossi et al., 2019).

Topographic images produced from CTs are especially useful for studying the head and neck regions of the body. Due to its ability to provide spatial and high contrast visualization of the anatomy of the head and neck, CTs are a major contributor to the advancement of minimally invasive surgical techniques, such as the insertion of a tube into the body through a process called endoscopy (Dammann, 2014). Additionally, CTs are the primary diagnostic imaging tool for Rhinosinusitis, which is a common disease of the paranasal sinuses. This imaging tool can track the areas of the sinuses that have not responded to treatment and indicate variants in the anatomy that may have contributed to the development of rhinosinusitis (Dammann, 2014). Moreover, CTs are the main X-ray imaging tools used for assessing head- and neck-related tumours, inflammatory diseases, analyzing the bones that surround the brain, evaluating hearing loss and planning for surgeries in the temporal region of the brain (Dammann, 2014). Furthermore, X-ray radiography has been used by dentists for decades and is still widely used to assess anomalies of the teeth, surrounding bones and tissues. X-ray radiography is the primary technique used to evaluate teeth- and gum-related diseases such as endodontal disease and periodontal disease. It can also provide an image of all the teeth in a patient's mouth, including the underlying bone, which allows for easy evaluation of a patient's dental health. It is able to assess inflammatory diseases of the mouth, analyze the jaws for any trauma and plan for dental implant surgeries (Dammann, 2014). Overall, CT and radiography are X-ray imaging technologies that have contributed to significant advancements and provided major contributions to the study and management of head- and neck-related conditions.

Chest X-rays are useful for diagnosing a variety of conditions such as cancer, pneumonia, tuberculosis, heart failure and pulmonary diseases. A chest radiograph is the most frequently used tool to analyze abnormalities of the lung. Additionally, CTs are able to detect artery blockages of the pulmonary artery to the lungs, pulmonary hypertension and damaged lung tissue (Wielpütz et al., 2014). More recently, chest X-rays or chest CTs have aided in the diagnosis, analysis of disease progression and detection of complications in coronavirus disease 2019 (COVID-19). COVID-19 is a viral infection characterized by respiratory distress and the global spread of COVID-19 has been recognized by the World Health Organization as a pandemic (Martínez Chamorro et al., 2021). Chest X-rays are the first tool used to assess whether a patient has SARS-CoV-2 because it is low-cost and readily available. Unfortunately, chest X-rays are less sensitive than CTs and may present false negatives due to the limitations of the technique. However, chest CTs are able to conduct quick tests and produce high-resolution images to evaluate the severity of COVID-19. Although chest CTs have a sensitivity of 97%, chest CTs have a low specificity of 25% for COVID-19 diagnoses (Martínez Chamorro et al., 2021). Despite the few limitations of chest X-rays and chest CTs, X-ray imaging technologies have been considered effective tools for the management of COVID-19. Overall, X-ray imaging technologies are regularly used to evaluate a variety of chest-related conditions and improvements in X-rays can aid in more sensitive imaging technologies and disease-specific analysis of diseases (Martínez Chamorro et al., 2021).

Limitations of X-ray imaging

Despite the value of X-ray imaging in diagnosing and monitoring diseases, the ability to accurately diagnose a condition may be unreliable. Chest X-rays are the most common form of X-ray imaging technique, yet they are ranked as the most difficult to interpret. For example, distinguishing between a chest X-ray of pneumonia compared to that of heart failure may be the most challenging for physicians (Satia et al., 2013). Additionally, 90% of errors in lung cancer diagnoses occur on chest X-rays. It is difficult for physicians to differentiate a lung lesion from bones, vessels and other structures when analyzing a chest radiograph. Anatomical structures such as the ribs, heart, blood vessels, diaphragm tend to overlap each other in a radiograph, which contributes to unnecessary anatomic noise and potentially hiding cancer-indicators (Del Cielo, 2017). In addition to radiography, lung cancer diagnosis also presents issues when using CTs to evaluate the lungs due to errors in tissue scanning, tumour recognition and decision making. When using X-ray imaging techniques, tumours located at the top of the lungs are missed 72% of the time and lesions located at the middle of the lungs are missed 65% of the time (Del Cielo, 2017). With lung cancer being the poorest surviving cancer and contributing to 1.59 million deaths worldwide, the inaccuracy of diagnosing lung cancer with X-ray imaging is devastating (Del Cielo, 2017).

Pneumonia is another example of a condition that is often misdiagnosed with chest X-rays. Radiography and CTs are the main methods of pneumonia diagnosis, but the features of pneumonia are often misinterpreted as symptoms of other similar lung conditions. This is especially harmful in lower income countries where X-ray imaging technologies and skilled clinicians are scarce. Since pneumonia affects 7% of the world's population, the inaccuracy of diagnosing pneumonia has detrimental outcomes (Hashmi et al., 2020).

Early diagnosis of lung cancer and pneumonia would drastically increase the patient survival rate. The use of artificial intelligence in conjunction with medical imaging may aid in the processing and interpretation of X-ray images (Hashmi et al., 2020). In recent years, machine learning algorithms have been established to more easily distinguish between additional structures in the anatomy of the lungs. There have been models created that were able to recognize pneumonia with a test accuracy of over 98% (Hashmi et al., 2020). Since computer-based imaging technologies are becoming popular, the use of this technique may be available at a low cost. It is estimated that further investigation in the use of artificial intelligence in the field of X-ray radiology may supplement clinical decision making and improve diagnosis accuracy of a variety of health conditions (Hashmi et al., 2020).

## Conclusion

X-rays have contributed to the development of X-ray imaging technologies such as radiography, fluoroscopy and CT. These technologies have made advancements in the diagnosis and treatment of various health conditions and are considered a valuable tool in the healthcare system. Considering their high global demand and importance in managing conditions such as rare diseases, X-rays should be further investigated to improve their imaging accuracy. X-rays may be developed with the addition of artificial intelligence to support clinicians.

**What Science is Involved in X-Rays?**

Introduction

Having examined the history and importance of nuclear imaging techniques of X-ray in modern medicine, it brings us to the inquiry of the scientific mechanics of modern X-ray imaging. This chapter aims to explain the science behind the medical application of X-rays, starting with an explanation of the definition of an X-ray, how its properties allows it to penetrate certain objects but not others, the film that is used, and how the two are used in combination in medical settings to create radiographs that aid in diagnosis and/or the plan of treatment. Furthermore, the health concerns involving X-ray induced mutations will be explained as well as its application in radiation therapy. This will be illustrated through a common analogy that will be explained in the next section.

General Overview

In an attempt to simplify the complex process that is the usage of X-rays as a medical imaging technique, an analogy will be used throughout this chapter to ground the explanations to the overall picture. The process of taking an X-ray is similar to shadow puppetry. There is a need for a light source such as a flashlight, a hand in front of the flashlight creating the shadows and lastly, the wall on which the light is reflected off, where applicable. In the case of the nuclear imaging technique, an X-ray source will act as the flashlight, the X-ray acts as the light, body parts act like the hand and a film will act as the wall. The details of this relationship will be explored further in this chapter. This analogy is illustrated below in figure [1].

Figure [1] Illustration of shadow puppetry. (Carr, et al., 2014)

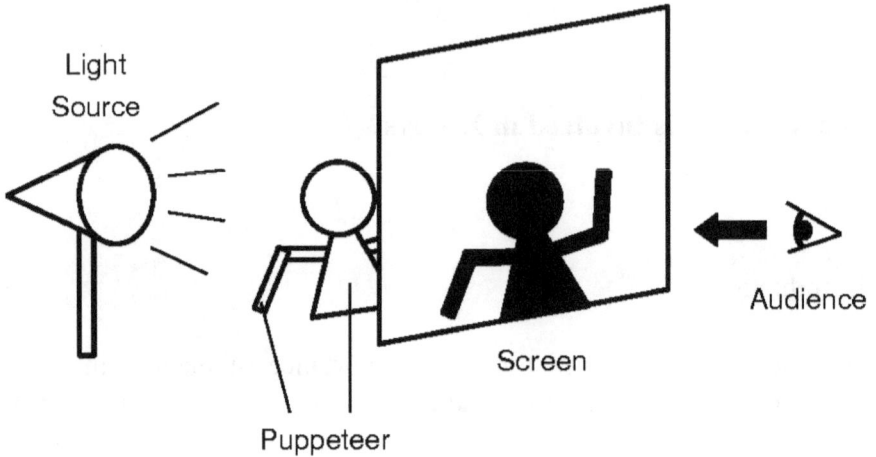

**Figure 2.1 – A shadow puppet theatre setup**

The Definition of X-ray

Similar to light, X-rays are a form of electromagnetic radiation (National Institute of Biomedical Imaging and Bioengineering, n.d.). In order to further examine the nature of X-rays, examination of electromagnetic radiation should be done first.

The term electromagnetic radiation describes cyclic repeating electromagnetic waves (Seibert, 2004). It includes radio waves, microwaves, infrared, visible light, ultra violet, x rays and gamma rays. At some wavelengths, it appears as visible light, at others it takes the form of radiation we cannot see, such as ultraviolet, infrared and X-rays. This wave can be described by both time and space using the terms: period, indicating the time, and wavelength, also denoted as $\lambda$, indicating the distance between repeating points of the wave (Seibert, 2004). The cycle has repeating units of the sinusoidal wave. The length of time that one cycle occurs for is referred to as the period, expressed in seconds (Seibert, 2004). Conversely, the number of cycles that occur per second is referred to as frequency (f), which is usually expressed in hertz (Hz) (Seibert, 2004). The relationship between the frequency and period is inversely equal, f=1/period (Seibert, 2004). Electromagnetic radiation travels at a constant velocity, C, also known as the speed of light, which is approximated at 3.0x10^8 m/s (Seibert, 2004). The relationship between velocity ( c ), wavelength (λ) and frequency (f) is denoted in the equation C = λ f (Seibert, 2004). The aspects of the electromagnetic wave are illustrated in the figure [2] below.

Figure [2] Labelled diagram of the electromagnetic wave. (Seibert, 2004)

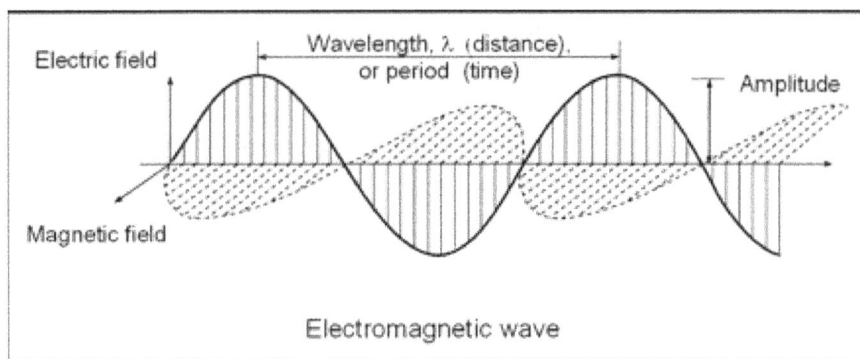

At higher energies and extremely short wavelengths, exhibited by X-rays, electromagnetic radiation exhibits particle-like traits (Seibert, 2004). This allows its photons, the smallest unit of electromagnetic radiation, to collide with matter (Seibert, 2004). This means that an x-ray photon with enough energy can remove electrons bound to an atom (Seibert, 2004). This process is known as ionization. This explains why X-rays are often referred to as ionizing radiation. It was discovered that photon energy is directly proportional to the frequency of the wave, meaning the higher the frequency, the higher energy (Seibert, 2004). This relationship is denoted as E=hf where E is energy and h is Planck's constant, 6.62 x 10^-34 J-s (Seibert, 2004). Energy in this equation is highlighted in Einstein's most famous equation of E=mc^2, also known as Einstein's theory of relativity in which E denotes energy, m denotes mass and c denotes the speed of light (Seibert, 2004). This states that energy is equal to the product of mass and the square of the speed of electromagnetic radiation. Einstein further states that mass and energy are interchangeable and that mass and energy are conserved in the interaction. This is a concept that is exemplified in nuclear medicine when a positive electron also known as a positron is emitted from proton rich unstable atom and interacts with a negative electron, resulting in an antimatter/matter conversion of the total mass of the positron and the electron into energy (Seibert, 2004). In radioactive elements with unstable nuclei, energy is released during the decay of the nucleus to a more stable, lower energy form. Therefore, when X-rays interact with matter, they collide with electrons. Sometimes the X-ray transfers all of its energy to the matter and gets absorbed. Other times, it only transfers some of its energy and the rest is scattered (Seibert, 2004). The frequency of these outcomes depends on how many electrons the X-rays are likely to hit. Collisions are more likely if a material is dense or if it's made of elements with higher atomic numbers like platinum and gold which means more electrons (Seibert, 2004). These power and ionization properties of X-rays permit its powerful application in the medical field. This will be elaborated in a later section in this chapter under medical application of X-rays. Prior to that, the production process of X-rays will be examined.

The Production of X-ray

In our analogy, a light source is needed to produce the light. In the case of X-rays, this is true as well. When Röntgen did his initial experiment, X-rays were produced when high energy electrons in a cathode tube hit a metal component, they either slowed down and released extra energy or kicked off electrons from the atoms they hit which trigger a reshuffling that again releases energy (Seibert, 2004). In both cases, the energy was emitted in the form of X-rays, a type of electromagnetic radiation with higher energy than visible light and lower energy than gamma rays (Seibert, 2004).

We can see through the demonstration of this experiment that X-rays result from the conversion of the kinetic energy attained by electrons accelerated under a potential difference into electromagnetic radiation (Seibert, 2004). As a result of that collisional and radiative interactions. An X-ray tube and X-ray generator are necessary for X-ray production and control (Seibert, 2004). The X-ray tube provides the proper environment and components to produce X-rays, whereas the X-ray generator provides the source of electrical voltage and user controls to energize the X-ray tube (Seibert, 2004). The basic components of an X-ray tube includes 2 electrodes, the cathode and anode 1-2 cm apart in a vacuum made of glass or metal. High voltage cables connect the cathode and the anode from the X-ray generator (Seibert, 2004). There also exists a separate isolated circuit that connects the cathode filament to a low voltage power source (Seibert, 2004). Both X-ray tubes and X-ray generators contribute to the production of X-rays.

X- ray Film

In the analogy used before, a wall was needed to display the shadow poppet. Where the light that was not able to pass through the hand, a shadow was left on the wall whereas where there was nothing blocking the path of the light, there would be light on the wall. The role of the wall is similar to that of the X-ray film. The film displays the radiographic image which is like the poppet shadow. Unlike the wall, the most common X-ray film consists of a couple of layers. An outer layer of cellulose triacetate or polyester encapsulating an emulsion of a silver halide such as silver bromide and gelatin (Møystad et al, 1996). When the silver halide in the emulsion is exposed to light or X-rays, it produces a silver ion and an electron (Møystad et al, 1996). The electrons then attach to the sensitivity speck which attach the silver ion which is positively charged (Møystad et al, 1996). When the silver ions attach and clump, metallic silver which is black in colour are formed. In areas which are not exposed to X-rays will stay light. Intermediate areas will be reflected in the intermediate shade range.

Medical Application of X-rays

As mentioned above, X-rays have very short wavelengths. In contrast to visible light, this gives it the ability to pass right through many kinds of matter (Seibert, 2004). This makes them powerful for many applications - most notably in the medical applications because they can make images of organs, like bones without harming them (National Institute of Biomedical Imaging and Bioengineering, n.d.). Since bones contain calcium, a denser material and a higher atomic number than most other tissues, they are able to stop some of the X-rays. This is what casts a light "shadow" on the dark film background. For example, if you place your hand in front of a flashlight, it will cast a shadow on the wall behind it. In a similar manner, X-rays directed at the body produce shadows on the opposite side that can be registered on film or on a digital sensor. As a result, the dimensional images formed by the X-rays are quite useful in revealing structures within the body and can easily display fractures in bones (Seibert, 2004). Bones are dense and full of calcium, which has a relatively high atomic number which makes them absorb X-rays whereas soft tissue on the other hand is not as dense and contains mostly lower atomic number elements like carbon, hydrogen and oxygen (National Institute of Biomedical Imaging and Bioengineering, n.d.). Therefore more of the X-rays penetrate tissues like lungs that will participate in darkening the film.

These 2D pictures are only useful up to a point. When x-rays travel through the body, they can only interact with the atoms along the path. To have a more holistic view of what's happening, you would have to take X-ray views from many angles around the body and use them to construct an internal image. This is what doctors do when they construct a computed tomography (CT) scan (Pogue & Wilson, 2018). A comparison can be made between an X-ray and a CT scan. With an X-ray, the density change due to a solid tumor in a patient is visible but the depth beneath the surface is unknown. If you take x-rays from multiple angles, you should be able to find the tumor's position and shape. This is essentially how a CT scanner works. The X-ray beam is rotated around the patient and often moved down the patient's body with the X-ray source tracing a spiral trajectory (Pogue & Wilson, 2018).The comparison between X-rays and other imaging techniques will be discussed in a later section.

Mutations due to X-rays

Although X-rays are quite safe, they still carry a small chance of causing mutations, most notably  reproductive organs and tissues like the thyroid. This is why lead aprons are often worn on areas not being imaged to block them  (National Institute of Biomedical Imaging and Bioengineering, n.d.). In high concentrations, X-rays and other types of high energy radiation can be used to destroy cells by damaging their DNA (National Institute of Biomedical Imaging and Bioengineering, n.d.). In a study conducted in 1998 by Thosmas et al, it was found that X-rays induced mutations in mouse embryonic stem cells (Thomas et al., 1998). This in vivo study, meaning a study on live organisms, was one of the first to map the exact mutations that were X-ray induced through lab techniques. The data showed that X-rays can be used to induce deletion complexes in mouse embryonic stem cells. The effects of mutations in the DNA of cells are random. Sometimes they can cause diseases, most often there's no effect. Without X-rays, all healthy bodies have a rate of mutations happening. The immune system will normally clear out the cells that have mutations that are affecting its function and its growth. The only time it's a concern is when the immune system misses a mutated cell and the mutation causes cells to reproduce at a higher rate. This is what is known as tumour. A small sized benign tumor can be removed surgically and can be harmless however when a benign tumor grows to a size which impedes functions of organs or tissues or mutates and metastasizes, meaning it travels throughout the body to replicate it becomes a larger medical concern. This is commonly known as cancer. X-rays can be a concern because they do induce mutations which is a factor to consider when choosing to do repeated X-rays.

Radiation therapy for the treatment of cancer

With the same pathway as healthy cells, the high energy radiation of X-rays can also eliminate cancerous cells. However, the radiation dose used for treating cancer is much higher than the radiation dose used for diagnostic imaging (National Institute of Biomedical Imaging and Bioengineering, n.d.). Therapeutic radiation can come from a machine outside of the body or from a radioactive material that is placed in the body, inside or near tumor cells or injected into the bloodstream (National Institute of Biomedical Imaging and Bioengineering, n.d.).

## Conclusion

In this chapter, the science involved in production and interaction of X-rays for the purposes of medical applications have been examined to show the intricacies of this technology that is used widely in western medicine. According to the Pan American Health Organisation (PAHO), it has been estimated that 3.6 billion diagnostic X-ray examinations have been conducted in hospitals and clinics each year worldwide.

# What Have We Learned About X-Rays in Recent Medical History?

Ever since its discovery in 1895, X-rays have been used non-invasively to acquire high-resolution images of thick biological tissues (Chen et al., 2012). In recent years, a number of novel contrast methodologies have been discovered, which have expanded X-ray biomedical applications into functional as well as structural imaging. A few of these techniques include X-ray absorption, X-ray fluorescence, and X-ray excited optical luminescence. These methods are useful for monitoring biological, molecular, and elemental components and entities in tissues, tumors, pharmaceuticals, and implants (Chen et al., 2012). Researchers expect that these techniques will radically develop our capacity to research in situ biochemistry, disease pathology and various other eminent diseases.

X-ray Imaging Contrast

One of the most recent advances in X-rays is the improvement of resolution and quality. The ability to distinguish two materials depends on how well their linear x-ray attenuation coefficient, $\mu$, is measured (Chen et al., 2012). As the number of atoms increases, X-ray attenuation increases, and as X-ray energy increases, the attenuation decreases. The conventional projection X-ray employs a polychromatic X-ray source to measure the average transmittance of the X-ray energies. Recent advances in energy dispersing cameras have allowed the amount of delivered X-ray energy to be measured at each pixel (Chen et al., 2012). With this technology, high-energy and high-spatial resolution can be applied to differentiating tissues according to their elemental compositions.

X-ray Fluorescence (XRF)

Real-time X-ray imaging, known as fluoroscopy, has been in clinical use since shortly after Röntgen presented his discovery of X-rays. Fluoroscopes were originally composed of an X-ray source and a fluorescent screen, which were placed opposite each other, allowing a patient to be placed in between (Gingold., 2014). As the remnant beam passed through the patient, it impinged on a fluorescent screen and resulted in a visible glow, which was detectable by the practitioner. In today's real-time X-ray imaging, fluorescent screens are combined with electronic devices that make the glowing light into a video signal that can be visualized on an electronic display (Gingold., 2014). Unlike most other forms of X-ray imaging, fluoroscopy is unique in that the images that are produced appear in real-time which can be used to evaluate dynamic biological processes and guide interventions. To clarify, real-time signal processing refers to the actual time during which a process or an event occurs. Digital fluoroscopy systems make this perception possible by capturing and displaying images at a rapid frame rate, typically 25 to 30 frames per second (Gingold., 2014). Many fluoroscopy systems utilize high voltage-generators and X-ray tubes that are similar in design and construction to those used for general radiographic purposes.

X-ray luminescence and X-ray fluorescence computed tomography (CT) are two emerging technologies in X-ray imaging that offer functional and molecular imaging capability. These two varieties of tomographic imaging modalities use external X-rays to excite secondary sources, either light or secondary X-rays, which are then obtained for tomographic reconstruction. These modalities excel beyond the limits of sensitivity in current X-ray imaging and are capable of extracting information from cellular tissues and organs by transforming X-ray imaging into molecular imaging (Ahmad et al., 2014). It is also possible to image three-dimensional elemental composition in specimens by using X-ray fluorescence computed tomography (XRCT). In this method of tomography, CT imaging is used in conjunction with secondary X-ray fluorescence measured from elements within a sample to create tomographic images (Chen et al., 2012). In XRCT, distributions are reconstructed from 2D representations to produce 3D maps. The XRCT method involves irradiating a sample with an external X-ray beam and detecting the X-ray fluorescence with an energy dispersive detector. Each X-ray fluorescence photon is quantified by counting the pair of electron/hole produced after absorption in the detector (Chen et al., 2012). As in conventional CT, the sensor is placed opposite the X-ray source in order to acquire the X-rays propagated through the subject. The detectors in XFCT, on the other hand, are positioned outside the beam to retrieve the X-rays being emitted by the subject (Chen et al., 2012).

By using synchrotron based micro X-ray fluorescence images, detection limits and spatial resolution compare favorably to conventional X-ray fluorescence (XRF) instruments. With XRF, high resolution images can be achieved for biopsies and cell studies. Synchrotron X-ray fluorescence investigations with hard X-rays, especially their depth of penetration, make them highly suitable for cellular studies (Chen et al., 2012). This method also has a major disadvantage since synchrotron XFCT imaging facilities are expensive and take up a lot of space (Ahmad et al., 2014).

64-slice CT available - 2004

A computerized tomography scan (CT) or computerized axial tomography scan (CAT) combines data from multiple X-rays to generate a precise image of internal structures within the body. In comparison to conventional X-rays which use a "stationary" scanner, the CT device rotates on a rotating gantry, emitting X-ray energy that is absorbed by the body (Cramer et al., 2018). Several technological innovations have been made since the CT scan's introduction in 1972, starting with a single layer of imaging to a multi-layer spiral with two, four, eight, and sixteen detectors. Most noticeable are the improvements in rotation speed resulting in an increase in both temporal resolution and spatial resolution along the z axis (Cademartiri et al., 2005). Clinical progress in x-ray imaging has been made in 2004 as a result of increased adoption and implementation of Multi-slice CT (MSCT) machines beyond 16 slices (Nikolaou et al., 2004). In each rotation, the scanner creates 64 anatomical images that are of high-resolution. This new method of 64SCT Cardiac Imaging proved beneficial in obtaining improved spatial resolution with an isotropic voxel size of 0.4mm and a temporal resolution of 83-165ms

(Nikolaou et al., 2004). During a 64-slice CT scan, a sequence of x-ray views taken from many different angles are combined, and then a computer is used to reconstruct the "slices" to produce cross sectional images of the body (U.S. Food and Drug Administration, 2020). The newer MSCT scanners are expected to offer much better visualization of the fine details. Diagnostic imaging of this kind is recommended when the soft tissues of the body - such as internal organs - are to be evaluated. In comparison to x-rays and CT scans, 64-slice CT scans are better able to give a higher quality image of the soft tissue structure size, shape, and location (Coshocton Regional Medical Center, 2021). Particularly for fast-moving vessels with detailed imaging, such as the heart and coronary arteries, these benefits hold great promise. Before the emergence of 64SCT-based imaging, the narrow dimensions of coronary arteries, along with their quick motion, made imaging very challenging. Recent generations of 16-slice and 64-slice CT have made detection of hemodynamically relevant coronary artery stenoses possible as it can produce an image of the heart and coronary vessels in fewer than five seconds without causing blurring of the image (Achenbach, 2006).

Dual-source CT - 2006

Taking diagnostic-quality images of the coronary arteries used to be a considerable challenge because of the constant motion of the heart and their small size (Johnson et al., 2008). Multidetector CT scanners and gated cardiac imaging provided a way to overcome many of the challenges of scanning the heart in motion. In many patients with heart rates above 65 beats per minute, obtaining diagnostic images with the 64-detector scanners was not possible due to deviations from myocardial motion. Additionally, patients with significant heart rate variability could not be imaged (Johnson et al., 2008). With 16- and 64-slice CT, beta-blockers are typically employed in order to decrease heart rate. However, this approach is time consuming and constrained by contraindications (Johnson et al., 2008). Since the introduction of dual-source CT scanners in 2006, these limitations are no longer present. The heart rate was no longer a limitation, enabling doctors to even closely examine patients who had arrhythmias or atrial fibrillation. A dual-source CT scanner consists of two x-ray tubes that are assembled at a 90 degree angle and additionally consists of two sets of detectors in the gantry (Petersilka et al., 2008). Reconstruction of an image can be accomplished through 180-degree projections when the gantry rotates only 90 degrees. In addition, the gantry rotation is much faster than with preceding scanners, resulting in 83-ms temporal resolution (Lin and Alessio, 2009). The fastest 64-slice multidetector CT scanners have a temporal resolution of 165 ms, which is slower than the gantry rotation (Miller et al., 2008). Furthermore, thorough cardiac assessments that include wall motion and valve function can be performed (Johnson et al., 2008). One potential advantage of the dual source CT system lies in the capacity of both x-ray tubes, which in theory work simultaneously, to produce x-ray photons of different energies. Thus, with this advantage, additional information can be learned about the object scanned.

Research on the accuracy of dual-source coronary CT angiography (CTA) has been completed in a few studies.

One study examined 109 patients with acute chest pain on a dual-source CT scanner. The images were evaluated for the cause of chest pain and the results were correlated with an invasive coronary angiography. From this study, it was found that overall sensitivity for the cause of chest pain was 98%, and the correlation with invasive angiography revealed 100% sensitivity (Johnson et al., 2007). According to another study conducted with 90 patients who had intermediate pretest probability for coronary artery disease, the sensitivity for detecting coronary stenoses over 75% was 100%, and that for half the stenoses was 88% (Leber et al., 2007). These studies provide evidence that dual-source coronary CTA is a robust and fast chest pain assessment tool.

First clinical 3D mammography exam performed in the US - 2011

X-ray imaging in mammography provides an early detection of cancer and other breast diseases through the examination of the breast (NIH, 2021). It is used as both a diagnostic and screening tool. Mammograms include putting a patient's breast on a flat support plate and compressing it with a parallel plate known as a paddle. In an x-ray machine a small burst of x-rays is produced which passes through the breast to a detector on the opposite side. The images produced are called mammograms. On a film mammogram, areas of low density tissues, such as fat, appear translucent (i.e darker shades of gray nearing the black background), whereas areas of dense tissue, such as connective glandular tissue or tumors, seem whiter on a gray background (NIH, 2021). Digital mammograms use the same x-rays as conventional mammograms, but they make use of solid-state detectors to record the x-ray pattern as the x-ray passes through the breast. These detectors transform the x-rays passing through them into electronic signals that go to a computer (NIH, 2021). Digital mammography has several advantages over film mammography, among them the ability to alter image contrast to improve clarity, the capability of using computer-assisted detection in order to identify abnormalities, and the ease of sending digital files to others for a second opinion. Further, digital mammograms may lead to fewer retakes and more importantly, lower radiation exposures (NIH, 2021). Digital Breast Tomosynthesis is also referred to as 3D mammography, and its availability and popularity have increased significantly since it was approved by the Food and Drug Administration (FDA) in 2011. This screening method is a form of breast cancer testing when x-rays of the breast are captured at different angles to generate thin cross-sections (NIH, 2021). The 3D representation is akin to the 3D images that the standard CT technology produces. Compared to CT, the radiation dose from Tomosynthesis is dramatically reduced as significantly fewer x-ray beams are projected through the breast as well as to the rest of the chest (NIH, 2021). Three dimensional mammography seems to identify breast cancers better than two dimensional mammography according to initial studies, but large-scale studies in randomized trials are still ongoing to compare tomosynthesis to two dimensional mammography (NIH, 2021). Because of this, researchers are unable to determine for certain whether 3D mammography will be better or worse than standard mammography in ensuring accurate results and identifying early breast cancers in all types of patients.

Because of its use of x-rays, intraoperative mammography exposes patients to ionizing radiation for a short period of time. Generally, women don't experience any risk from receiving regular mammograms, as the benefits outweigh the radiation exposure (NIH, 2021).

Risks related to this dose appear to be higher among younger women (under the age of 40). However, in some situations, mammography may be more beneficial than radiation exposure when used to detect breast cancer in women under 40. It is possible, for example, that a mammogram will reveal a benign mass, so that it does not need to be treated (NIH, 2021). The surgeon may also be able to remove a tumor that is malignant and detected early by mammogram, before the tumor spreads and requires more aggressive treatments such as chemotherapy (NIH, 2021).

<u>First ever XACT image and Field Resurgence - 2013</u>

Over the past century, X-ray imaging has proven invaluable for medical diagnosis and nondestructive testing (NDT). Still, two serious challenges remain: radiation harm and inaccessibility of the sample (Samant et al., 2020). X-ray imaging is required to be balanced against the risk of high doses of ionizing radiation applied to healthy tissues due to the fact that ionizing radiation is itself a carcinogen. The limitation of inaccessibility of sample derives from the fact that x-ray absorption based images require placement of the x-ray detector opposite from the source. In cases where it is impossible to place a detector in the direct path of x-ray photons, this can limit the applications of x-ray imaging. Recently introduced in 2013, x-ray induced acoustic computed tomography (XACT) offers a novel imaging modality (Samant et al., 2020). With this technology, high X-ray absorption contrast is incorporated into the technology with the advantages of 3D propagation provided by high-resolution ultrasound waves (Li, 2020). In XACT, x-ray induced excitation causes localized heating and thermoelastic expansion. This results in detectable ultrasonic emission, which allows imaging. Through the use of XACT, it is possible to access low-dose, fast, 3D imaging requiring only single side access. This is the first time that an x-ray induced acoustic signal was applied to tomographic imaging (Samant et al., 2020).

From the 1950s, when X-ray induced acoustic (XA) waves were observed, XA imaging has been used to image the distribution of sample tissues that are absorptive to X-rays. The principle of XA imaging is similar to that of laser based photoacoustic (PA) imaging. A carefully selected wavelength of excitation light allows PA imaging to provide a wealth of anatomical, functional, and molecular information about biological tissues. Since X-rays have straighter paths than optical light, which is used in PA imaging, the energy is able to penetrate the body deeper (Lee et al., 2020). Photon-electron interactions are the mechanism by which X-rays can produce ultrasound. Photoelectrons, Auger electrons, or electromagnetic radiation in the form of fluorescence is generated when X-rays strike the sample and excite the inner-shell electrons. All three of these add thermal energy to the sample as fluorescence reabsorption, Auger electron absorption, and photoelectron absorption. Through electron-phonon interactions, these processes ultimately result in excited electrons depositing kinetic energy into the sample. This energy causes a very small (mK) temperature rise in the sample, resulting in a subsequent thermal expansion and the emission of an acoustic wave in the ultrasonic frequency range (Samant et al., 2020).

A study conducted by Tang and colleagues investigated the feasibility of x-ray-induced acoustic computed tomography (XACT) for early breast un-palpable microcalcification detection in three dimensions. This method provides a precise and accurate 3D image of breast volume that overcomes the disadvantage of tissue superposition in mammography (Tang et al., 2018).

Conclusion

In recent medical history, there has been significant development in our understanding and application of x-ray imaging. Over the past couple of years, the faster, easier, and safer applications of x-ray imaging has been attributed to the newer x-ray imaging models that have been refined and upgraded from older imaging techniques. These newer models have helped make diagnosis of profound diseases such as breast cancer easier, with observable changes in quality, accuracy, and processing speed. As our understanding of x-ray imaging as a diagnostic and screening tool continues to expand, it paves way for many more revolutionary findings.

# What Safety Concerns Arise When Using X-Ray Imaging?

Exposure to ionizing radiation is practically inevitable since its use in health care during diagnosis, therapies, and medical screening programs is ever so increasingly embedded in modern medicine. Likewise to every aspect of patient care and treatment, X-ray imaging, which utilises ionizing radiation to generate images of the insides of one's body, is bound to be associated with risks and safety concerns. While medical exposure to radiation is generally low in dosage, it is the single most significant source of radiation exposure to the general public compared to other natural sources in many countries (Averbeck et al., 2018). Radiation exposure in X-ray technicians, workers in the nuclear industry, as well as the general public (to a much lower extent of course) is unavoidable in a society where the use of nuclear energy and ionizing radiation has become indispensable. It is crucial to make light of potential human and environmental health risks associated with X-ray imaging in order to enforce the necessary safety restrictions as well as grant society unlimited access to the beneficial uses of ionizing radiation. However, it is important to note that considerable ambiguities still remain in the science which analyzes the health effects resulting from exposure to low levels of radiation emitted by X-ray technologies, despite decades of research. Thus, more research is still necessary to determine the mechanisms involved in low dose radiation exposure and their potential safety implications.

Doses of ionizing radiation from medical imaging technologies are constantly being scrutinized in medical and lay press due to recent research papers which highlight the increased cancer risks associated with computed tomography (CT) scans (Lin, 2010). Berrington de Gonzalez estimates that nearly 2% of annual diagnosed cancers may be related to CT scans performed in the United States during 2007;

data that is comparable to Brenner Hall's estimates of 1.5% to 2% (Lin, 2010). Cancer is medically defined as a disease where normal cellular functions and activities are altered resulting in abnormal cell proliferation and growth (Averbeck et al., 2018). Abnormal metabolic activity is generally induced by internal or external stress factors which may result in gene mutations, such as those in tumour suppressor genes, which are involved in regulating cell growth, as well as oncogenes which modulate the cell cycle and cell division (Averbeck et al., 2018). Since cancer pathology is generally based on genetic mutations, one may correctly postulate that DNA is the primary cellular target site for radiation-induced stochastic effects, like cancer (Averbeck et al., 2018). Radiation-induced cancers typically involve mutations and damage including DNA double stranded breaks, where both DNA strands are severed, and clustered DNA lesions, where a DNA site has been damaged by multiple different mechanisms (Averbeck et al., 2018). The quantity and degree of damage is directly proportional to the quality of the radiation and the dosage (Averbeck et al., 2018). Human cells have repair mechanisms which aim to fix DNA damage. Sometimes, these mechanisms are faulty resulting in unrepaired or mis-repaired DNA lesions, which ultimately can lead to genetic instability, and thus cancer (Averbeck et al., 2018).

Ionizing radiation and its causation in cancer has been a significant topic of research, and thus experimented substantially. In many studies, strains of mice with different DNA repair capacities were treated with fractionated low-dose radiation (100 mSv daily). All mice exposed to radiation expressed an increase in DNA damage in the lung cells (bronchial and alveolar), heart cells (cardiomyocytes), and brain cells (cortical neurons) of mice (Averbeck et al., 2018). Strains of mice which had more compromised DNA repair systems yielded a significant increase in persistent pathological foci, ultimately pointing to an increased risk of carcinogenesis, the initiation of cancer formation (Averbeck et al., 2018). Radiation exposed stem cells have also been studied and researched further in an effort to analyse potential risks to fertility and development (Averbeck et al., 2018). Fractionated exposure to radiation (from 10 mSv to 100 mSv) also resulted in persistent DNA damage in spermatogonia stem cells in mice - thereby suggesting a potential, although controversial, possible reproductive risk for humans via out of target radiation exposure during medical imaging or radiation therapy (Averbeck et al., 2018). X-rays expose patients and sometimes technicians to a low dose of radiation (defined as 100 mSv) as well as low dose-rates (those defined as less than 6 mSv/h) (Averbeck et al., 2018). Thus, these sorts of studies, where mice are exposed to low doses of radiation, may allow scientific and medical researchers to postulate the potential physiological effects on humans. However, their findings are not factually indicative of low dose radiation effects in humans. Thus, further medical trials or epidemiological studies on atomic bomb survivors may prove prevalent in furthering research on the health risks associated with radiation.

Long-term cohort studies on the Japanese survivors of the catastrophic 1945 atomic bombing of Hiroshima and Nagasaki provides robust data which analyzes the health implications of radiation - albeit at significantly larger levels compared to those emitted by medical imaging technologies. These cohort studies categorize long term effects into leukemia, solid cancer, and noncancer health effects (Douple et al., 2011). Leukemia deaths were the first observed radiation-associated long-term health risks (Douple et al., 2011). Of the cohort survivors, there were 315 deaths associated with leukemia, of which 45% are estimated to be attributable to radiation exposure above 5 mSv (Douple et al., 2011). This proportion of mortality to leukemia increases in survivors exposed to greater doses of radiation, 86% of which died due to exposure greater than 1000 mSv (Douple et al., 2011). Contrarily, of the cohort survivors' solid cancer incidents, only 11% have been attributed to radiation exposure greater than 5 mSv (Douple et al., 2011). This attribution proportion increases with greater doses of radiation, such that for survivors exposed to at least 1000 mSv, 48% of the solid cancer cases are the result of exposure to the atomic bomb's radiation (Douple et al., 2011). In terms of noncancer health effects, lens opacities, thyroid diseases, cardiovascular and other late-onset diseases are all observed and associated with some form of radiation exposure (Douple et al., 2011). Within 3 to 4 years after the bombings, survivors who were exposed to high doses of radiation developed eye cataracts (Douple et al., 2011). Additionally, the prevalence of hyperparathyroidism increases directly with radiation dose starting at 1000 mSv (Douple et al., 2011).

From a cardiovascular standpoint, reported physiological changes associated with radiation include elevations in systolic and diastolic blood pressure and cholesterol levels, as well as arch calcification, hypertension, and ischemic heart diseases. However, the risk for these sorts of cardiovascular diseases are undefined at radiation levels below 500 mSV (Douple et al., 2011). Evidently, there is clear evidence of a cancer risk at radiation doses above 100 mSv. However, the radiation-induced cancer risk between 10 and 100 mSv is still controversial. While this range is of little relevance to medical imaging such as X-rays due to their miniscule radiation emission, it may be of importance for many high dose medical imaging technologies such as CT, nuclear cardiology, and other complex interventional radiology and cardiology procedures involving fluoroscopy since they typically fall into this controversial dosage range, and potentially past if repeated many times (Lin, 2010).

It is quite evident that over exposure to X-ray radiation imposes adverse effects on living cells and organisms which ultimately may result in many life-threatening ailments. Thus, it can be asserted that functional abnormalities may still be caused by low levels of radiation exposure (Meo et al., 2006). A study conducted by a group of medical doctors and scientists at the King Khalid University Hospital, Saudi Arabia, studied the effects of X-ray radiation on human immune cells, specifically on the total leukocyte count (TLC) and the phagocytic activity (a process where immune cells engulf other cells or particles in an immune response) of Polymorphonuclear neutrophils (PMN) (Meo et al., 2006) . Forty-two healthy X-ray technicians with an age range of 25-50 years were recruited and comparatively matched with another group of forty-two healthy volunteer subjects in terms of their sex, age, and ethnic variation for statistically unbiased results. A Beckman Coulter counter was performed to determine TLC (Meo et al., 2006). On the other hand, the phagocytic activity of PMN was determined via quantifying the chemiluminescence (CL) response. The results suggest that while the X-ray technicians exhibited a relatively low level of exposure to radiation (2.896 mSv per year), their mean value of the CL response for their PMNs significantly decreased (p<0.0005) compared to their controls (Meo et al., 2006). However, no significant difference was observed in TLC counts between both of the two groups, regardless of the levels of exposure between them (Meo et al., 2006). Based on these results, it may be deduced that exposure to X-ray radiation reduces the physiological functions and capacity of PMNs even at relatively low levels of radiation exposure. Ultimately, this implicates prolonged or heightened radiation exposure to potential adverse immunological symptoms (Meo et al., 2006).

Periapical X-rays are an important diagnostic tool in dentistry as they permit the visualization of a patient's teeth and the detection of medical abnormalities (Nejaim et al., 2015). The radiation doses required for periapical radiographs are below the range which produce health concerns. However, such imaging technology requires the use of lead foils, which generally function to absorb residual X-ray beams, which may result in the environment pollution of lead when not disposed of properly(1,7). Lead waste can be maintained in the topsoil for nearly 2000 years, where it waits to be uptaken by plants, a clean water supply, and consequently by humans and animals alike (CDC, 2018). Lead is a neurotoxin, so prolonged exposure in humans may result in lead poisoning. Symptoms may include anemia, weakness, kidney damage, as well as many other adverse long term effects, including death (CDC, 2018). The Eco Dentistry Association (EDA) estimates that American dental offices dispose of 48 million lead foils yearly. Despite such a significant environmental and health threat, 60% of dental offices in the United States have yet to eliminate the use of traditional X-rays. Digital X-rays replace the need for lead foils and thus prevent potential adversities associated with traditional periapical X-rays, however, they have yet to be adopted by the majority of dental clinics in the United States.

On another note, traditional X-rays require further film processing such that in order for scanned images to be visualized, they must be chemically cured. The sheer amount of water used to rinse the film, the chemicals applied, as well as the fumes released during the curing process all result in major environmental implications, such as the contamination of water supplies and the environment with harmful photo chemicals (Hewitt, 1993). Exposure to these chemicals in radiographers and darkroom technicians, who often engage in processing X-ray films, experience an array of symptoms (often coined "darkroom disease") such as: sinus pains, sore eyes, blurred visions, aching and ringing ears, chest and ear infections, weight loss, tiredness, dry skin, inflamed nostrils, and nausea as a result of exposure (Hewitt, 1993). For instance, the fixing process in film processing, which is responsible for stabilizing the final image and removing the unexposed silver halide which results in the reduced metallic silver that visualizes the image, uses a photographic fixer solution containing toxic chemicals (Hewitt, 1993). Some of which include Ammonium Thiocyanate and Boric anhydride, all of which are known to irritate the skin, eyes, and the respiratory tract (Hewitt, 1993). They may also be toxic to the thyroid, kidneys, liver, and blood (Hewitt, 1993). Chemical inhalation and ingestion are thus hazardous because they can induce organ damage (Hewitt, 1993). This safety risk can once again be resolved with the use of digital X-rays since they feed directly onto a computer. So printing the film no longer requires an environmentally hazardous process and thus avoids potential health risks involved in the fixing process.

While some risks concerning radiation exposure from medical imaging technologies may be ambiguous, undefined, or controversial, it is imperative to reduce patient radiation exposure unless absolutely necessary. Physicians should review a patient's imaging history before medical imaging technologies are considered. This may reveal a history of a high cumulative radiation exposure, and thus this may negate the need for more imaging or at least promote a lower radiation dose alternative (Lin, 2010). Often, this is not the case as many patients experiencing chronic or recurrent conditions, such as renal colic disease, had an average cumulative dose of greater than 50 mSv from medical imaging over the span of three years (Lin, 2010). Physicians should hesitate when prescribing medical imaging with moderate to high radiation doses for young and pregnant patients, as well as others with a high body mass index (BMI) (Lin, 2010). Although radiation risk to a fetus from medical imaging technologies is low, they should be avoided until absolutely necessary. This is because potential biological effects of high dose radiation exposure in utero may include childhood cancer, prenatal death, intrauterine growth restriction, small head size, mental retardation, and organ malformation (Lin, 2010). On the other hand, younger patients have a higher risk from radiation due to the fact that they have more remaining years of life where radiation induced cancer may develop (Lin, 2010). In a study by Smith-Bindman et al, it was estimated that for a 20 year old patient, their risk for radiation-induced cancer is twice that of a 40 year old patient (Lin, 2010). Additionally, young children are 3 to 4 times more radiosensitive than adults and so lower dosage exposure poses greater risks (Lin, 2010). Since the thickness of the region being imaged increases for patients with a higher BMI, greater X-ray penetration is required to produce a necessary quality of image (Lin, 2010). If the same amount of radiation is used for patients of both low and high BMI, the image for the higher BMI patient will appear rather grainy or noisy compared to that of the lower BMI (Lin, 2010). Thus, certain techniques for reducing radiation exposure in medical imaging technologies are not possible for high BMI patients. Ultimately, such patients will be subjected to higher doses of radiation, which typically also have a higher association to cancer (Lin, 2010).

Physicians should refer to radiologists concerning questions about when certain medical imaging technologies are applicable for their patient, especially if they are attempting to reduce radiation exposure. For instance, compared to a standard CT, multiphasic CTs produce better detection and characterization of liver lesions, since the same organ can be scanned up to 4 times (Lin, 2010). However, the radiation dose of multiphasic CT is almost 4 times that of the single phase CT dose (Lin, 2010). On the other hand, magnetic resonance imaging (MRI) can produce greater diagnostic accuracy than multiphasic CT, and potentially with a lower radiation dose. For certain patients and cases, an MRI may be a better option. Radiologists are thus a good source of resource for questions concerning different types of medical imaging technologies as well as those concerning the minimization of radiation dose. New technologies have also been introduced to minimize radiation exposure such as the adaptive statistical iterative reconstruction (ASIR) technique for CT scans (Lin, 2010). ASIR creates a less noisy image, thereby allowing the radiation dose to be decreased significantly (Lin, 2010). For instance, using ASIR in conjunction with other radiation reducing techniques can reduce a 20 mSv radiation dose to that of only 1 mSv (Lin, 2010). To put this into perspective, a patient will be exposed to approximately 20 mSv of radiation when prescribed a CT coronary angiography, and because of this exposure, the patient now has a 1 in 270 chance of developing radiation-induced cancer (Lin, 2010).

Undoubtedly, reducing exposure to radiation with the use of new technological improvements can drastically curtail a patient's risk for radiation-induced cancers.

If radiation can cause cancer and other chronic health risks, does that suggest one must avoid radiation at all costs? To put it simply, exposure to radiation is unavoidable because humans are all exposed to some degree of natural radiation from the sun, the earth, radon in homes, and cosmic radiation during long-distance flights (Komaroff, 2013). In fact, the average American is exposed to 3.7 mSv of natural background radiation per year (Komaroff, 2013). Comparatively, a chest X-ray, with two views, would expose a person to 0.01 mSv (Komaroff, 2013). So the radiation emitted from one chest X-ray would equate to nearly the same amount of natural background radiation a person would be exposed to on a single average day (Komaroff, 2013). Dental X-rays result in a much lower radiation exposure than chest X-rays. So the potential health risks associated with radiation are most likely not associated with exposure to dental X-rays. The same goes with airport scanners since they emit such little radiation that in order to surpass the natural background radiation one usually is exposed to per year, one must pass through more than 25,000 scans (Komaroff, 2013). However, not all X-ray technologies emit low radiation, for example, CT scans can emit up to 4-7 mSV for chest imaging. Moreover, this average may be higher for CT abdominal imaging (Komaroff, 2013). The American College of Radiology recommends no more than 100 mSv lifetime diagnostic radiation exposure - this equates to nearly 10,000 chest X-rays but only 25 chest CTs (Komaroff, 2013). Thus, hesitancies and worries towards basic low dose X-rays are essentially futile because the current research suggests that there are zero to no health risks associated with such a low dosage. On the other hand, while other medical technologies like CT can emit rather significant levels of radiation, exposure to them once in a lifetime is not exactly a certain determinant of radiation induced cancers. However, it is still necessary, both for doctors and patients, to be wary of when and how often such medical imaging technologies (that emit high dose radiations) are used because radiation effects are ultimately cumulative. Thus, as long as doctors and patients attempt to reduce exposure to radiation and maintain a cumulative exposure below a 100 mSv threshold, either by consulting radiologists, ensuring medical imaging facilities are updated with the latest radiation reducing technologies, and discriminating when medical imaging is unnecessary, cancer and other health risks are ostensibly nullified.

## What are Opposing or Alternative Imaging Technologies to X-rays?

While X-ray scans are often used in diagnostics and research, different imaging technologies may be required on a case by case basis. Different injuries or medical concerns may incline physicians to opt for one imaging technology over another. Additionally, imaging technologies can vary in terms of invasiveness, detail provided, areas of the body possible to image, and speed, making the choice of which imaging technology to use an important one. Health care providers and researchers alike may need an imaging technology to supply structural information or to image a specific area of the body. Regardless of research, diagnostic, or monitoring requirements for each case, a patient themselves may request an alternative imaging technology if feasible. Some patients may have concerns about invasiveness, claustrophobia, or fears about unnecessary exposure to chemicals or radiation. For these reasons and more, advances in technology in the medical field have given health care providers their pick of diagnostic imaging tools, such as MRI, fMRI, CT, ultrasound, PET, and many others, each with their own advantages and disadvantages.

As mentioned earlier, different imaging technologies are sometimes required depending on the area of the body needing imaging. One imaging technology may be used more frequently than another when scanning a specific body part or structure because of differences in contrast or detail. To provide an example, X-ray scans are not typically used for brain imaging because of how complex the structure of the brain is; an X-ray scan is not capable of capturing the intricacies and density of the sulci—the folds of the brain (Breedlove & Watson, 2017, pg. 50). To combat this, a technique called angiography was developed to modify X-ray scans by injecting an X-ray blocking dye, like those discussed in Chapter 4, into the brain vessels to enhance contrast in X-ray imaging and thereby providing more detail (Breedlove & Watson, 2017, pg. 51). The additional detail allows for improved imaging accuracy and ease of diagnosis or monitoring. The brain is a particularly difficult organ to image because of its numerous ridges and folds, so angiography is not the only technique or imaging technology developed that can image the brain with precision.

Magnetic Resonance Imaging (MRI) is an imaging technology used in research and the diagnosis and monitoring of injuries or medical conditions. It is different from X-ray scans in that no X-rays are involved—instead, radio waves on the opposite end of the electromagnetic spectrum are used in combination with a powerful magnetic field to produce images (Mayo Clinic, 2019). The purpose of the magnetic field is to realign the water molecules inside the patient's body, allowing them to release signals when targeted by radio waves; those signals are then captured in a series of images which depict a cross-section of the patient's body (Mayo Clinic, 2019). The magnetic field is generated by a high-powered magnet shaped like a tube, inside of which the patient lays down for the procedure (Mayo Clinic, 2019). The confined space may instill a sense of panic in patients suffering from claustrophobia, so a mild sedative may be given to help them relax (Mayo Clinic, 2019). The magnetic field then moves the water molecules in the patient's body by forcing the protons inside the nuclei of those water molecules to line up parallel to each other (Breedlove & Watson, 2017, pg. 53). The protons are disrupted when the radio waves hit them, but return to the random configurations they were originally in once the radio wave pulse is shut off (Breedlove & Watson, 2017, pg. 53). As they return to their original orientations, they emit radio frequencies back, which are then picked up by detectors and combined by a computer (Breedlove & Watson, 2017, pg. 53). In brain scans, the images generated by the computer depict a map of the brain's density, because the frequencies of the radio waves emitted by the protons differ at different densities (Breedlove & Watson, 2017, pg. 53). The computer can even use the images to create a 3D map of the brain (Mayo Clinic, 2019). The use of a contrast agent, which is injected via IV, is sometimes required in order to increase contrast in the resulting scan (Mayo Clinic, 2019). MRI scans are more detailed than those produced by an X-ray scan, which can make diagnosis and the monitoring of medical conditions easier for physicians and other medical care providers (Mayo Clinic, 2019). Another benefit of MRI scans is that they use radio waves instead of X-rays, which have the potential to be harmful (Breedlove & Watson, 2017, pg. 52). Although MRI scans are a generally safe diagnostic test, there is a risk factor for patients with metal inside of their bodies, like those with pacemakers, metal screws, plates, or shrapnel, or intrauterine devices (IUDs) (Mayo Clinic, 2019). Because MRI scans require the use of a powerful magnet, there is always the potential risk of the metal on or inside a patient's body being attracted by the magnet, which is why patients are asked to remove any metal jewelry, clothing, or accessories prior to the procedure; however, regardless of whether attraction occurs, metal can affect the image quality of an MRI scan (Mayo Clinic, 2019). Similarly, there is a minor risk for patients with tattoos—some types of tattoo ink, particularly the darker ones, may contain metal (Mayo Clinic, 2019). Pregnancy is also a risk factor because research is unclear about the impact of magnetic fields on fetuses; even women postpartum that are breastfeeding should speak with their physicians, because MRI scans may involve the injection of contrast agents (Mayo Clinic, 2019). Lastly, the use of contrast agents can also affect those with kidney or liver problems, so this should be discussed with one's physician (Mayo Clinic, 2019).

MRI scans are the most commonly used imaging technology for brain and spinal cord imaging, and can be used for diagnosing things like aneurysms, tumors, strokes, traumatic brain injuries, and spinal cord disorders (Mayo Clinic, 2019). Moreover, MRI scans are able to show small changes in the brain, like the wearing down of myelin (Breedlove & Watson, 2017, pg. 53). Myelin is the fatty substance surrounding an axon—the end of the neuron that carries away signals—which increases the speed of those signals (Breedlove & Watson, 2017, pg. 35). As a result, MRI scans can aid health care providers in diagnosing multiple sclerosis (MS), because the wasting away of myelin is a major component of MS (Breedlove & Watson, 2017, pg. 53). There is also a more specialized type of MRI called the functional MRI—or fMRI—which produces sharp images in a sensible amount of time (Breedlove & Watson, 2017, pg. 53). The fMRI scan is used to image the flow of blood in the brain, allowing researchers and medical care providers to determine what parts of the brain are most active when performing specific tasks, such as taking in visual stimuli or manipulating an object (Breedlove & Watson, 2017, pg. 53). The fMRI works similarly to an MRI—high-powered magnetic fields are used to track miniscule changes in brain metabolism, or the usage of oxygen, in different parts of the brain (Breedlove & Watson, 2017, pg. 53). The areas of activity highlighted in fMRI scans can be used to piece together how the brain works and how different areas of the brain network to carry out cognitive processes (Breedlove & Watson, 2017, pg. 53). Being able to track brain activity and con- nections in the brain is incredibly helpful for scientific research and the monitoring of medical conditions. Furthermore, fMRI scans can assist medical care providers in evaluating the level of damage to the brain sustained during a head injury or from other disorders, or be used to identify the critical brain structures involved in language and movement in brain surgery candidates (Mayo Clinic, 2019).

A health care provider can request an MRI scan for a wide range of body parts or structures, not just the brain. One example is the heart and blood vessels—an MRI scan can allow a physician to determine the damage caused by heart attacks or heart disease, or if there are any blockages within the blood vessels (Mayo Clinic, 2019). Addi- tionally, an MRI scan can provide structural information, like the size of the chambers of the heart, or the thickness of the heart's walls (Mayo Clinic, 2019). Another example is the bones and joints—an MRI scan might reveal bone tumors and infections, or joint and disc abnormalities (Mayo Clinic, 2019). Other organs, like the uterus or ovaries, the liver, the pancreas, and the kidneys, may also be scanned via MRI (Mayo Clinic, 2019).

Another imaging technology used in the medical field and in research is computerized tomography (CT). A CT scan also involves the use of X-rays, but the technique used is slightly different from that of an X-ray scan. In a CT scan, X-rays are directed at the body while im- ages are taken at different angles—these images are then combined mathematically using computer technology to form a detailed image of the bones and tissue (Mayo Clinic, 2020-a). A CT scan works by measuring the amount of X-rays absorbed by the body, which re- veals the density of the tissue the X-rays are directed at (Breedlove

& Watson, 2017, pg. 52). One benefit of a CT scan is that it is more detailed than an X-ray scan is, however, as a consequence of the extra detail, a CT scan exposes the patient to more radiation than an X-ray scan would (Mayo Clinic, 2020-a). This is because of the amount of X-ray exposure required to take the additional images (Mayo Clinic, 2020-a). Naturally, the increase in radiation exposure may prove worrisome to patients, but research done on CT scans has not shown any long-term harm (Mayo Clinic 2020-a), as described in previous chapters. Another drawback of CT scans is that they are less detailed than MRI scans, as they are only a medium resolution imaging technology (Breedlove & Watson, 2017, pg. 52). CT scans can be used for diagnosis, like when identifying bone fractures, clots, or tumors, or for monitoring medical conditions and the progress of prescribed treatments (Mayo Clinic, 2020-a).

Ultrasound scans, also referred to as sonography, are another type of imaging technology that are frequently used in the medical field (Mayo Clinic, 2020-c). As the name suggests, ultrasound scans use sound waves at high frequencies to produce images of the body (Mayo Clinic, 2020-c). Sound waves are less harmful than X-rays, but are also less efficient at travelling through bone or gas, making ultrasound scans less detailed and less effective at imaging certain body parts than X-ray, CT, or MRI scans (Mayo Clinic, 2020-c). For this reason, ultrasound scans are not typically used for imaging the head or the lungs (Mayo Clinic, 2020-c), as the lungs are filled with air and the head is largely bone due to the skull. Despite this, ultrasound scans are a safe and proficient method of imaging the body, which is why they are so commonplace. During an ultrasound, a water-based gel is applied to the skin to help reduce air pockets, which are difficult for sound waves to travel through (Mayo Clinic, 2020-c). A transducer—a small device used to emit sound waves—is then pressed over the skin to capture any sound waves that bounce back from the body (Mayo Clinic, 2020-c). Those collected sound waves are then electronically converted into an image by a computer (Mayo Clinic, 2020-c), much like the other imaging technologies previously described. Although most ultrasounds are done outside of the body, ultrasounds can sometimes require the insertion of the transducer via a probe into an opening in the body, like for transrectal or transvaginal ultrasounds (Mayo Clinic, 2020-c). The openings used are natural openings in the body—the sonographer does not create an opening for the probe (Mayo Clinic, 2020-c). Ultrasound scans can be used to assess things like pregnancy, blood flow, breast lumps, and genital or prostate issues (Mayo Clinic, 2020-c).

A positron emission tomography scan, or PET scan for short, is an alternative imaging technology to X-ray scans. It is unlike an X-ray scan because it utilizes a radioactive agent, also called a tracer, in order to image the body and its organs (Mayo Clinic, 2020-b). The tracer can be injected, ingested, or inhaled, depending on which organ or body structure is being imaged (Mayo Clinic, 2020-b). The purpose behind the tracer is that it collects in areas within the body that undergo high chemical activity, which coincide with disease, and the tracer build-up lights up on the scan (Mayo Clinic, 2020-b). A PET scan is a valuable tool in monitoring and diagnosing medical conditions, particularly cancers or heart disease (Mayo Clinic, 2020-b). The reason PET scans are used to detect cancer is because cancer cells are high in chemical activity, or metabolic rate; despite this, some cancers may not appear in a PET scan, and some areas of the body may light up that do not necessarily signify cancer (Mayo Clinic, 2020-b). Some cancerous tumors that do appear include: lung, brain, thyroid, prostate, and melanoma, among others (Mayo Clinic, 2020-b). For heart disease, a PET scan can capture the areas of the heart which have the least amount of blood flow, which might suggest a clogged artery (Mayo Clinic, 2020-b). The procedure of a PET scan is much like that of an MRI scan—the patient lays down in an enclosed tube that is full of radiation detectors which track the tracer's location in the body (Breedlove & Watson, 2017, pg. 53). The scanner may induce claustrophobia in some patients, but mild sedatives can be provided if necessary (Mayo Clinic, 2020-b). One advantage of PET scans is that they can potentially reveal medical conditions that other imaging technologies are not yet able to (Mayo Clinic, 2020-b). PET scans can also be used to research brain activity—one method utilizes radioactive glucose which is administered while the participant completes a task; the areas of the brain primarily responsible for executing the task will use up the radioactive glucose and light up on the PET scan (Breedlove & Watson, 2017, pg. 53). The scans, alongside mathematical techniques, can be used to create a metabolic map which depicts what areas of the brain contribute to different tasks (Breedlove & Watson, pg. 53). Although there is little risk involved in a PET scan, the risks that do exist are similar to those of an MRI scan—pregnant and breastfeeding women are cautioned that their child may be exposed to minor radiation from ingestion of the tracer (Mayo Clinic, 2020-b). PET scans can also be used in conjunction with MRI or CT scans so health care providers have more information for diagnosis (Mayo Clinic, 2020-b).

To conclude, there are many imaging technologies that can be used as an alternative to X-ray scans. The reasons why a physician or researcher may utilize one imaging technology over another are numerous, because of how different each imaging technology is. Some are greater than others at imaging specific body structures, like the brain, or are faster than others. Some carry more potential risks, like radiation, and some are invasive while others are not. While X-ray scans are a commonplace imaging technology with an array of benefits, the wide range of alternative imaging technologies available only serves to improve research, the diagnosis and monitoring of medical conditions, and to ease the work of health care providers.

## What Misinformation or Conspiracy Theories Exist Regarding X-Rays?

Conspiracy theories and misinformation have long plagued the scientific community, positing that menacing motivations underpin anything from medical technologies to therapeutic treatment. Following the massive growth of social media, there has been a dubious amplification of deceptive information targeting the ordinary users. The scope of this problem is greatly neglected—in a study of over 120,000 electronic news articles, researchers found that misinformation and falsehoods far surpassed the identification of trustworthy evidence (Goertzel, 2010). A non-exhaustive list of compelling factors increasing the rejection of medical sciences include (Fazio et al., 2015):

**Availability heuristic:** This mental shortcut relies on readily recalled information regardless of its fallacy. The alarming nature of medico-scientific false claims allow for faster recollection hence increasing their patent significance.

**Illusory truth phenomenon:** Following repeated exposure, there is an innate tendency to accept false information although our intellectual capacity tells one otherwise. This reiteration effect causes the perception of familiarity to subdue rationality influencing our logical deductive reasoning.

**Fallacy of anecdotal vividness:** The personal testimonials influencing the emotional impulses are more agreeable than dull narratives. Accordingly, when coupled with an ostensive personal claim, a groundless health assertion acquires increased traction.

It is facile yet misguided to dismiss such misconceptions as a sector of a small cohort consisting of undetectable eccentrics, however, this merely minimizes the challenge. Mere exposure to medical myths can contort the unsuspecting party's judgement leading to a global communal detriment. Compliance with medical technology is not simply binary; the varying factors implicated in risk-assessment generate complex probing of definite wagering agents. The radiative properties of X-rays have resulted in a marked rise of misinformation across the scientific community. It is widely accepted that the biochemical carriers of communication in human physiology are subject to mutations with age and exposure to radiative output. However, it is important to acknowledge the existence of multiple identifiable bioinformation systems prone to gradual breakdown accompanying clinical immersion (Tomà et al., 2017). In other words, the correlative properties of one constituent leading to a disease does not necessarily signify an identical response by other factors. The notion of native background radiation is not naturally accepted by patients, families, or physicians. There stands a sentimental perception of risk associated with fear, anxiety and concern. When assessing the risks and benefits, the risk of skipping a diagnostic test leading to wrongful diagnosis is often overlooked.

Over the years, a variety of claims have garnered considerable traction, including the assertion that X-rays are highly carcinogenic and can trigger chronic health implications. Although classified as a carcinogen by the World Health Organization (WHO), one meta-analysis involving 15 developed countries revealed a minimal 0.6-0.18% increase in cancer risk from diagnostic X-rays (de Gonzalez & Darby, 2004). Most of the harmful exposure is due to CT scanning and nuclear imaging requiring larger radiation doses than traditional X-rays. For example, a chest CT delivers 70 times as much radiation as a chest X-ray. Such misconceptions arise from an improper employment of the linear non-threshold (LNT) model based on over-simplification of disease emergence (Andronikou, 2017). Adopted in the 1950s, the LNT assumption is a dose-response model estimating stochastic health effects following radiation exposure (Doss, 2018). Stochastic health effects are cases of spontaneous emergence with a proportional probability to dosage yet independent severity. The LNT model presumes a linear correspondence between stochastic health risks and dosage with no reduction in thresholds (Baumann et al., 2020). Ultimately, LNT suggests that the sum of multiple minor exposures equates to a single large exposure causing an equally probable stochastic effect. Contrastingly, radiation-induced effects caused by skin damage encapsulate deterministic health effects not measured by the LNT model. Given the epidemiological support of its application, the LNT is a common model used to calculate the probability of radiation-induced illnesses yet, controversially, low dose relevance provides lower predictive statistical confidence. Essentially, it follows a linear relationship insinuating that radiation simply leads to mutations, and mutations lead to cancer. However, this model is easily invalidated as cancer incidence does not correlate linearly with abrupt mutations. Most of scientific knowledge on ionizing radiation risks originates from long-term analyses of the survivors from the 1945 catastrophic bombing of Hiroshima and Nagasaki. Such studies tracked over 25,000 victims that received radiation equal to three or more CT scans (< 50 mSv) (Harvard Health Publishing, 2020). The appearance of excess cancers following the immediate exposure triggered international alarm with reference to radiation hazards. However, as it is often unacknowledged in the media, the atomic blast is not a complete model comparable to X-ray diagnostics. The key distinction stems from the simultaneous radiation unveiling of the nuclear bombs as compared to the gradual exposure of radiodiagnostic procedures. Such historical incidences catalyzed the development of faulty misconceptions regarding paralleled maturation of cancer when exposed to X-rays. Nevertheless, such incidents were founded on high-dose radiation independent from the low-dose radiation emitted by X-rays.

An alternative justification premised on the misuse of the LNT model suggests considering that children have a higher proportion of dividing cells, as such, they are thereby more susceptible to mutations caused by radiation (Andronikou, 2017). Although children remain more sensitive to radiation during their peeking growth period, diagnostics exams are often adjusted for smaller body sizes in order to evade overexposure. Such arguments are speculative and ignore the biological defences activated by low-dose radiation as the probability of DNA damage following low dose radiation exposure is extremely small. Following detection, temporary adaptive protection responses corresponding to alterations in gene expression egress (Feinendegen et al., 2004).

Some responses include:

**Detoxification of reactive oxygen species**
Radical detoxification is stimulated reaching an optimum capacity at 4 hours following irradiation and lasting up to weeks depending on cell types. Low dosage of radiation increases a superoxide enzyme parallelled with decreased oxidative degradation of lipids that lasts for weeks. Elevated glutathione involved in killer cellular activity is also observed in spleen tissue.

**Enhanced rate of DNA repair**
High dose radiation protection observed in human lymphocytes through chromosome aberrations reaches optimum capacity at 4 hours following irradiation. Such genetically-determined and cell-dependant protection remains effective against varying DNA damaging agents. This response enhances DNA repair rate by several folds.

**Removal of damaged cells**
Signal-induced cell death (apoptosis) of damaged cells occur within a few hours of high dose radiation exposure. Low dose radiation induces apoptosis followed by replacement of healthy tissue when a threshold of DNA damage is met. Alternatively, defective cells exit through premature maturation to natural aging systems.

**Stimulated immune response**
*In vivo*, damaged cells can be removed through induction caused by low dose immune response. A reduction in oncological metastasis and higher circulation of cytotoxic lymphocytes are also observed.

The unified operation of these defensive feedback can last up to weeks and be assayed by cellular reactions against regenerated irradiation (Feinendegen et al., 2004). The continuous and abundant physiological response operates against endogenous non-radiogenic DNA damage. The counteraction amid protective measures and damage favor protection at low radiative doses and damage at high dosage. Bystander effects analogous with high-dose radiated cells to neighboring units include a miscellaneous damage and protective measures. Contrary to the widely noted misbelief, risks of health effect determinations are insufficient at radiation dosages under 50-100 mSv (Andronikou, 2017).

Since the 1950s, there has been speculation that the biological outcome of normal aging is comparable to that of ionizing radiation exposure (Bertell, 1977). The fundamental misunderstanding arises from the much overlooked remark that X-ray radiation radiates emissions many magnitudes less than that of other radio-emitting technologies. X-rays do not induce extensive cellular damage nor destroy tissue membranes; instead they often do not influence the biological systems. Nevertheless, highly-radiating technologies introduce cellular error manifested through medical conditions posing a deficit in homeostasis restoration and sensing of internal conditions leading to symptoms similar to aging. What has previously been reported as a uniform radiation effect is rather dose-dependent.

Misconceptions concerning the teratogenic risk associated with diagnostic imaging have long been overestimated, leading to unnecessary termination of pregnancies (Ratnapalan et al., 2008). Pregnant women who had undertaken radiographic testing had a perceived major malformation risk of 26% whereas women exposed to non teratogenic agents had a lower (16%) perception risk. Such high notions of risk lead to unnecessary distress and delays in needed care. One study revealed that 10% of pregnant participants exposed to X-rays chose to terminate their pregnancy resulting from anxiety caused by widespread misinformation (Cohen-Kerem et. al, 2006). Additionally, women often opt out from working in the endovascular surgical neuroradiology field due to concerns of occupational exposure to ionizing radiation. Such anxiety is amplified by the known outcome of nuclear adversities and the continuous influx in medical literature regarding the long-term consequences of high dose radiation on fertility. Studies have proven that *in utero* exposure to ionizing radiation is dose-dependent with an unambiguous boundary. There are no indications that radiodiagnostic doses of ionizing radiation during pregnancy increase the incidence of intrauterine growth retardation or gross congenital malformations. Developing fetuses in the eighth to fifteenth weeks of gestation are, however, particularly sensitive to ionizing radiation. A millisievert (mSv) is assigned as the cumulative background radiation dose that an individual absorbs in each. Exposure to 1 milligray (mGy) produces 1 mSv dosage of radiation. Fetal exposure to <200 mGy — a dose much higher than diagnostic exposure — could induce fetal cancers, neural deficits, or miscarriages (De Santis et al., 2005). Such risk has been overblown by social networks thus stimulating the long speculated misconception of fetal damage. Radiation doses below 100 mSv pose negligible risks for complication advancements. Most X-ray scans such as dental, chest, arms, or legs, would not expose the direct X-ray beam to reproductive organs as additional lead aprons are worn for further protective measures (Tobah, 2020). When exposed to ionizing radiation emitted by X-rays, many pregnant women opt out to wear 2 or 3 aprons. Studies have demonstrated that using two 0.5 mm lead aprons resulted in embryo exposure to radiation less than a detectable minimum. Accordingly, guidelines set by the National Council on Radiation Protection and Measurements estimate that the radiation dose is minimalistic at 1.5 mSv for 9 months when a lead apron is worn (De Santis et al., 2005). The risk of harm to the baby is determined by the gestational age of the fetus and the amount of exposed radiation. A miscarriage might occur if a woman is exposed to extremely high-dose radiation in the first two weeks after conception. These dose levels, however, are not used in diagnostic imaging (ie. X-rays). Two to eight weeks following conception, high-dose radiation can increase the risk of fetal growth defects or birth restrictions. At eight to sixteen weeks, the risk of developing an intellectual or learning ability increases. Conversely, the standard dose of a single diagnostic X-ray exposure is much lower than the extreme doses affiliated with these ramifications. Individuals uncertain of such common misconceptions are encouraged to talk to their physicians prior to exposure to X-rays. Depending on circumstances, it can be possible to delay X-ray exams or adjust the radiation quantities.

Popular news networks and social platforms have accelerated the circulation of the distorted truths following inadequate extrapolation from literature data. Policy makers have long scrambled to content with impeding misinformation, notably in relation to medical reports (Chou & Klein, 2018). The broadcast of deceptive information is magnified by the echo chamber effect in which members of online networks are persistently echoed back reinforcements of their beliefs following reduced publicity of others' viewpoints. Increasing data has displayed an accelerated dispersal of falsehoods in comparison to factual evidence. These influences are exceedingly difficult to un cover while records on effective interventions to reduce conspiracies remain limited. Some productive approaches might include (Stojanov & Halberstadt, 2020):

**Increasing understanding:** Systematic doubt propels conspirative ideologies which can be neutralized by increasing understanding thus leaving the irrational ideas less probable.

**Instilling a sense of control:** Agreement of conspirative ideas can be the result of reduced control of an individual's environment. Improvement of one's overall awareness of control is a possible solution to deflating acceptances of conspiracy theories.

**Expert intervention:** Seeking physicians and medical experts to help better understand an overview of scientific topics are critical. A specialist's recommendation of X-ray testing has significant reassurance impact on parental decisions as well.

Considering the prevalence of these allegations, repudiations of such claims are imperative. Medical professionals seek all possible measures to limit radiation risk during testing. Physicians are at the frontline of attempts to raise awareness about the falsity and risk of these widely held beliefs. Efficient techniques are needed to reduce their impact. Nonetheless, simply providing information is rarely enough to persuade patients given the significant proportion motivated by misinformation. This is particularly true in the case of X-rays and medical conspiracy theories as a whole, because victims may feel additional vulnerability as their concerns seem unheard. With radiation, evidence suggests a psychological disposition to accept conspiracy theories despite clear objections by medical experts. Frequently, individuals develop a cognitive bias proclaiming to understand the full scope of a subject without recognizing their prejudice — a prime example of the Dunning-Kruger effect (Vandergriendt, 2020). A combination of organic growth of long-established misinformation and deliberative conspiracy theories have dominated the supremacy of misrepresentation of scientific findings. The role of social media in propagating contemporary fibs have long been exhibited throughout history. Similar to claims that X-rays are bio-weapons, Russian state forces commemorated fables about the hazards of 5G. Another element often forgotten is the augmentation of falsehoods by public figures, politicians, and celebrities.

Moving forward, here are five pointers to help you terminate misinformation:

## 1) Skepticism

Readers must be skeptical of all new information, whether it's from social media or the news. Inquire about evidence supporting a source's claims and ask for their deductive techniques. Compare such information with different outlets and verify the ubiquity of the information.

## 2) Misinformation Landscape

Misinformation is not a novel concept. However, the media outlets presenting such information are increasingly evolving and expanding their impact in the journalism scope. Such models depend on user engagement and are therefore not financially obligated to present factual information. Minimizing reliance on social media platforms for news assessment is recommended.

## 3) Emotional and Divisive Topics

Misinformation is the most effective topic targeting increased traction. Emotionally-driven news are often susceptible to bias with little factual information backing up said claims. This is an immensely employed approach in the development of X-ray conspiracy theories targeting pregnant women.

## 4) Investigation

Ask questions regarding the context of the addressed news: Is the material sponsored by a corporation, a politician, or another inherently skewed source? Is there sufficient evidence? Are the referenced resources reliable?

## 5) Ceasing Dispersal

Although it is important to correct a fault, often mending a misinformation is delicate. Clarifying the origin of a misconception by providing factual evidence is needed to persuade a deviation from false knowledge. When approaching social media influencers, news outlets, or family members, it is important to inform them of the flawed reasoning behind their claims while bearing in mind the delicacy of the situation.

# Where is X-Ray Research Headed in the Future?

## Introduction

The field of X-ray research covers a large variety of topics from recognizing problems with existing procedures to incorporating new technologies like artificial intelligence (AI) into X-ray usage. With the changes in our understanding of X-rays, X-ray research has also expanded to include evaluations of environmental concerns as well as the role of X-rays in hospital settings. The future of X-ray research can be discussed from two main perspectives: research focused on investigating current issues with X-ray technology and usage, and research to develop innovative improvements to expand the capabilities and applications of X-rays. The current issues perspective represents a reflective and corrective route for X-ray research. Whereas the innovative technology perspective represents more prospective and interdisciplinary approaches to investigating X-rays. By looking at prevalent topics of study in both of these perspectives, trends in future X-ray research interest can be observed.

## Current Issues

### *Radiation Exposure Safety*

As an important diagnostic tool and medical imaging technique, one of the main focuses of X-ray research is investigating existing safety concerns with X-ray usage. The most prominent safety issue with X-rays is ionizing radiation exposure (refer to Chapter 7). Over the years, an article by Harvard Health Publishing (2020) states there has been an observed increase in radiation exposure, which experts suggest is due to medical imaging including methods such as X-rays and CT scans. There are natural sources of ionizing radiation which people are exposed to regularly, but the higher dose radiation exposures from these imaging methods raises the key concern of increased risk of developing cancer (carcinogenic) (Harvard Health Publishing, 2020). However, the chance of this risk is not expected to be large, but the effects of radiation have only been observed since it's discovery in the late 1890s (Harvard Health Publishing, 2020). There has thus been a general pressure to reduce use of radiation exposure by medical imaging out of fear, which has been intensified by mainstream media perceptions of X-ray linkage to risk of cancer (refer to Chapter 9?) (Oakley & Harrison, 2021).

A study conducted by Oakley & Harrison (2021) observes the lack of significant effectiveness in five types of current efforts to reduce X-ray radiation exposure: campaigns promoting radiation reduction, lower-dose practices, alternative imaging methods (refer to Chapter 8), emphasizing carcinogenic risks to patients, and physicians reducing total number of exposures based on patient history. This study identified that all of these efforts shared the linear no-threshold (LNT) assumption (Oakley & Harrison, 2021). In an earlier paper by Oakley & Harrison (2020), the LNT concept is described through 'As Low As Reasonably Achievable' (ALARA), which was an acronym used in radiology to promote using X-rays only when necessary. There have been cases of reluctance from physicians to give or patients to receive X-ray examinations as a result, despite little current scientific evidence to support this ideology which has resulted in it's gradual abandonment (Oakley & Harrison, 2020). Oakley & Harrison's (2021) study concluded that current X-ray procedures are within safe limits based on our current understanding, and alternative procedures could create other risks. These two studies highlight the necessity and regular practice of evaluation of existing assumptions to examine whether these continue to be true to current understandings in the field of X-rays.

These results indicate a clear direction that the risk of radiation exposure was previously overestimated under the LNT model, which led to inappropriate avoidance of X-ray usage. Other methods have been used to assess the safety risks of X-rays. In a review on the necessity of spinal X-rays in the chiropractic profession, the dosage of the X-ray was compared to be lower than a years worth of background ionizing radiation exposure, and thus the benefits of conducting the spinal X-rays as a diagnostic tool outweighed the harms of the exposure risks (Jenkins et al., 2018). However, in a commentary by Sykes (2020), there is recognition for evidence against LNT at lower doses meaning less need for radiophobia for X-rays, but also raises that rejection of LNT is not possible due to insufficient current evidence to disprove LNT completely. It is still debatable as to what degree of an ALARA type approach should be maintained in regulatory systems before a more appropriate model is found to decide radiation dosages based on risk. This stems from no proven direct harm to the public by using this type of ideology as it can be interpreted as an overly conservative approach rather than a wrong approach.

Therefore, it is important that future research continues to investigate how to assess the true risk level of ionizing radiation exposure to make the appropriate adjustments to X-ray usage based on the principle of 'doing no harm'. This sentiment is echoed in the proceedings of the International Commission of Radiological Protection (ICRP)'s Fifth International Symposium on The System of Radiological Protection (2020) which states the current recommendation for radiological procedures published in 2007 took a decade of continuous revision to be established, and future symposia should continue to focus on this review and refinement process. This statement emphasizes the continuous need for rigorous and long experimentation processes to assess the radiation exposure risks of X-rays and thus make appropriate recommendations.

### *Protective Measures*
This opens the question of what protective measures are currently being used against radiation exposure, and how effective these methods are. Radioprotective or 'lead' shields have been used since the 1950s to protect against radiation exposure for risks of cancer, with specific shielding for gonads (reproductive organs) to prevent effect on future generations (NCRP, 2021). There has been a reversal in the attitudes towards using protective shielding due to the improvements in X-ray technology in the past 70 years as well as discoveries on the adverse effects of using these shields (ACR, 2021). The American Association of Physicists in Medicine (2019) and the National Council on Radiation Protection and Measurements (2021)

released statements explaining the rationale for this change against 'protective' shielding. Both highlight that since the 1950s, there has been an overestimate of risk of damage to reproductive functions as learned from the studies providing evidence against the ALARA ideology, and X-ray technology has improved to reduce the radiation dosage absorbed by 95% (AAPM, 2019; NCRP, 2021). Another key argument against using shielding is that the current shields does not decrease the scatter of radiation to the gonads and may not even fully cover the gonads due to variation in location by person, but actually interferes with the imaging technique by blocking important locations or interfering with the automatic exposure control functions of modern X-rays (AAPM, 2019; NCRP, 2021). This is an example showing the importance of studies on the effectiveness of a technology being repeated over long durations of time. With new understandings and improvements in the X-ray field, the flaws in this protective measure were discovered and are now being fixed. As new protective measures arise to replace gonadal protective shielding, they will also need to undergo these rigorous assessments to determine whether they are necessary and safe to put into effect. From these studies on radiation exposures, research on X-ray safety can be predicted to head towards determining whether protective gear is needed and if there are better solutions to lower radiation doses to reduce carcinogenic and other health risks.

### *Regulatory Research*

Following the overall rise in concern for safety issues with X-rays, research has also been conducted to adjust safety standards and regulations for their usage. In a report on a Canada wide survey on radiation regulations conducted by Bjarnason et al. (2019), it states that Canadian facilities have been generally following the changes in ICRP and NCRP guidelines on dosage constraints and shielding designs to protect occupational users and patients, but the specifics of each province vary in following the Safety Code 35 (which lists a current comprehensive guide to radiology regulation). Some provinces do not have regulation for radiation safety, and there are many provinces who are still using an older version of the dosage constraint guidelines (Bjarnason et al., 2019). This report by Bjarnason et al. (2019) recognizes that there is criticism for these guidelines, but also that without updating and establishing a synchronized regulation according to Safety Code 35 as endorsed by the Canadian Organization of Medical Physicists (COMP), this puts patients and physicians at risk when hoping they will set their own constraints. This is supported by evidence of a lack of routine badging in certain provinces to keep track of worker exposure to radiation, and similar levels of radiation exposure observed in x-ray radiation and nuclear medicine workers (Bjarnason et al., 2019).

Similar to observing the degree of appropriateness of X-ray usage and trends in misuse, continuing the complementary research on developing and assessing the following of appropriate regulatory procedures is essential to maintaining the safety of both worker and patient users. As highlighted in Health Quality Ontario's (2016) report on modernizing Ontario's radiation protection legislation, an extensive research process by an interdisciplinary expert panel is required to evaluate what changes can increase the benefit over harm in using equipment with radiation exposure risks. The balance of both radiation safety and regulation research focuses is integral for future X-ray research as new technological improvements and interdisciplinary collaborations bring new risks and challenges to user safety.

**Innovative Improvements**

Another key aspect of X-ray research is the developing advancements on X-ray technology to overcome previous barriers. With the evolution of X-ray technology, improvements such as color imaging and increased precision have arisen.

***Expanding on X-ray Capabilities and Contexts***
A main new expansion on X-ray capabilities is flexible detectors, which allow X-rays to image curved 3D objects compared to the normal flat panels of X-rays (Dumé, 2021). The key importance of 3D models is it's important to surgical planning and evaluation post-operation (Kasten et al., 2020). These detectors are made of semiconductive heavy metals which can operate and several bended angles without reduced resolution (Liang et al., 2020). This is made possible by the mechanism of persistent radioluminescence, which still has many flaws such as higher dosage and longer exposure time which still need to be refined (Dumé, 2021).

X-ray research has also begun to expand beyond technique improvement to environmental contexts. At Florida State University, a team of researchers developed flexible X-ray detectors which can be created at a lower cost and cause less harm to the environment (Florida State University, 2020). They found a material called organic manganese (II) hybrid $(C_{38}H_{34}P_2)MnBr_4$ which provides high performance results when used in X-rays, and it's creation can be easily done at room temperature which reduces facility costs (Xu et al., 2020). This material is also cheaper and lead-free, making it more eco-friendly and attractive than many other existing commercial options (Xu et al., 2020). With increasing attention to the environment, commercially viable and less harmful options are starting to be developed, and can be expected to be of increasing focus in future X-ray research.

3D color X-rays are another improvement which have only been made available in clinical trials in the past few years, and has been an innovative improvement in the visualization and precision of medical imaging (Gall, 2020). These improvements are evident in images shared by MARS Bioimaging using the new Medipix3 detector chips developed at CERN, which shows high definition images with no need for use of contrast agents to see blood vessels (Gall, 2020). This technique has created breakthroughs in detecting microfractures in bones, and has potential in being developed to detect blockages in soft tissues like heart blood vessels (University of Maryland Baltimore County, 2019). As this color technology advances with further research, it may be possible to increase the precision of X-rays to view on a more micro-level and identify issues in more tissues.

With the development of virtual reality (VR) technology, medical schools have begun to try incorporating this technology into medical education simulations. One example is a study by O'Connor et al. (2021) which observes first-year radiology students using a VR suite to simulate conducting an X-ray examination. The majority of students in this study reported increased confidence in conducting X-ray procedures to practicing with VR, and would advocate for more of its use in formative assessments (O'Connor et al., 2021). This is an example of mixing the experience of X-rays and VR technologies to create lower risk and more comfortable learning environments for students. However, there are still flaws such as glitches and differences with real patient experience feedback which still need to be advanced in future studies (O'Connor et al., 2021).

### *Artificial Intelligence and X-rays*
Besides VR technology, another recent area of interest is the use of artificial intelligence (AI) in combination with X-rays. The main AI technique being used in radiology right now is convolutional neural networks (CNN) which is a deep learning and machine learning technique which allows the X-ray machine to learn from several layers of data without explicitly coding for this (Mun et al., 2021).

AI has mainly been used to improve the initial interpretations or diagnostics of X-ray images. An example of this is in a study by Kasten et al. (2020), which produced fast and accurate results in 3D reconstructions of knee bones using X-ray images. By training machines to process X-rays to create 3D models, this reduces the radiation exposure risks of having to use higher dosage methods such as CT and MRI scans (Kasten et al., 2020). Future studies can take this technique and project them onto more areas of bone reconstruction. Another example is the application of AI to chest X-ray (CXR) interpretations. The IBM Research-Almaden lab worked with an interdisciplinary team of radiologists, machine learning researchers, project managers, etc. to build a trained algorithm which has no significant difference in accuracy between the machine and radiology residents in interpreting CXRs (Syeda-Mahmood, 2020). This shapes AI deep learning models as a supportive tool for radiologists and physicians, especially when radiologists are not always available (Bai, 2019; Syeda-Mahmood, 2020).

The U.S. Food and Drug Administration (FDA) recently approved clinical use of Critical Care Suite, an AI algorithm trained to work with X-rays at an accuracy of 96% in identifying pneumothorax (collapsed lungs, a severe condition) (Bai, 2019). This is an innovative tool as this condition can be easily overlooked even though it can be treated quickly, and can be incorporated into existing portable X-rays machines (Bai, 2019). The future of AI with X-rays can be expected to work towards more of these severe conditions, and improving its accuracy and depth of detail it can provide in helping interpret X-rays.

## The Influence of COVID-19

Coronavirus disease 2019 (COVID-19, the disease name, SARS-COV-2, the virus name) has pushed forward an innovative wave in X-ray research. As a pandemic that has dominated a major focus in the research of 2020 and 2021, many adaptations have been re-searched and implemented to X-rays to support this new COVID-19 context.

CNN has been heavily applied in research on chest X-rays (CXRs) to help identify COVID-19. CXRs have been one way to help iden-tify COVID-19 early on since it can induce pneumonia symptoms which can show patterns when observing the lungs even when a PCR test has given a negative (Balbi et al., 2020). In Ippolito et al.'s (2020) study, a radiologist with more than 15 years of experi-ence had a 57% accuracy rate for identifying SARS-COV-2 through CXRs. In a model trained by Borkowski et al. (2020), there was an overall accuracy level of around 93% in identifying and differentiat-ing COVID-19 pneumonia, non-COVID-19 pneumonia, and normal lungs. Although it seems that CNN has been able to greatly improve prediction of COVID-19 through identification of SARS-COV-2 symptoms in observing the lungs in CXRs, there is a factor of these studies being new in protocol and limited in number which needs to be critically assessed for bias. This would be a necessary direction for further study for this to be implemented more systematically into supporting COVID-19 diagnosis.

Portable X-rays are another important innovation in the context of this pandemic. In 2019, Health Canada and the FDA both approved an affordable portable X-ray solution to be used (Business Wire, 2020). CXRs have been preferred over using CT during COVID-19 because of relative cleaning inefficiency and the lack of availabil-ity, especially with it's aforementioned benefit in early diagnosis (Jacobi et al., 2020).However, the key benefit of this compact solu-tion extends to protecting medical workers as a safer imaging tech-nique for medical workers. By modifying existing portable X-rays, clear images can be taken through glass doors or walls at a 6-foot distance with no increased radiation risk to suspected COVID-19 patients (Gange et al., 2020; Mozdy, 2020). From a X-ray technologists perspective, this method protects them from long exposures with suspected COVID-19 patients which preserves their protective gear such as masks, and overall use response was positive due to no decrease in effectivity (Gange et al., 2020). This was a new application of an existing developing improvement on X-rays, and this may be an area for further investigation in improving worker safety during X-rays. Portable X-rays is still a field with limited literature exploring its effects on specific populations, and Toppenberg et al. (2020) identified that future research could be more targeted towards filling gaps in known patient effectiveness, satisfaction, and costs in lower-income and broader populations.

**Conclusion**

The future of X-ray research is headed towards creating safer and more effective X-ray examinations through improving upon previous misconceptions and incorporating new technology to expand what can be done with X-rays. This also means that research on medical X-rays is expanding to become more interdisciplinary with considerations in contexts such as education and hospital operations. With the continuing fast development of different technologies, X-ray research can be expected to improve through a variety of new solutions, which may follow with a set of new concepts and challenges to be evaluated.

# References

**Chapter 1:**

Bradford, Isabella. "Mending Broken Bones." *Two Nerdy History Girls*, 2014, https://twonerdyhistorygirls.blogspot.com/2014/02/mending-broken-bones-c-1769.html. Accessed 3 May 2021.

drzezo. "History of Fracture Treatment." *Musculoskeletal Key*, 2019, https://musculoskeletalkey.com/the-history-of-fracture-treatment/?fbclid=IwAR0c3Ah050SbHet5PPQLavREgypkeEfivTmxandeKjuvmoVyuzMnsT6TI_A. Accessed 3 May 2021.

JMVH Editorial Board. "History of Tuberculosis Part 1 - Phthisis, consumption and the White Plague." *jmvh.org*, 2021, https://www.britannica.com/science/history-of-medicine/Traditional-medicine-and-surgery-in-Asia. Accessed 3 May 2021.

Tsoucalas, Gregory. "Hippocrates, on the Infection of the Lower Respiratory Tract among the General Population in Ancient Greece." *Loongdom*, 2016, https://www.longdom.org/open-access/hippocrates-on-the-infection-of-the-lower-respiratory-tract-among-the-generalpopulation-in-ancient-greece-2327-5146-1000272.pdf?fbclid=IwAR3KeY_Klg54ZQ_pn4f1cnx5tVDYa6EKyIj8KwiI3tabVmuzAZ3qvM80ooo. Accessed 3 May 2021.

Upenn. "An Injured Limb in Early 19th Century America." *Exploring Illness Across Time and Place*, 2021, https://www.sas.upenn.edu/~rogert/arm19.html?fbclid=IwAR3cyJ3ni_GQIux3VZOKSwBFdNgzZcpP121e-DR5cTI6S1maBzokCSv6NKhw. Accessed 3 May 2021.

van der Mark, Wiel. "A Broken Leg in the Year 1350: Treatment and Prognosis." *exarc.net*, 2016, https://exarc.net/issue-2016-2/int/broken-leg-year-1350-treatment-and-prognosis?fbclid=IwAR0HB_qhoWkbfApUjf1fRY7mftniP9P_Jc7TIvqx2wrPoZamn8sUzs2GrEY. Accessed 3 May 2021.

## Chapter 2:

APS physics. (n.d.). *This Month in Physics History*. American Physical Society. https://www.aps.org/publications/apsnews/features/history.cfm.

A&E Television Networks. (2010, February 9). *Marie and Pierre Curie isolate radium*. History.com. https://www.history.com/this-day-in-history/curies-isolate-radium.

A&E Television Networks. (2009, November 24). *Scientist Discovers X-rays*. History.com. https://www.history.com/this-day-in-history/german-scientist-discovers-x-rays.

Bell, D. J. (n.d.). *William D Coolidge: Radiology Reference Article*. Radiopaedia Blog RSS. https://radiopaedia.org/articles/william-d-coolidge.

Columbia Surgery. (n.d.). *History of Medicine: Dr. Roentgen's Accidental X-Rays*. Columbia University Department of Surgery. https://columbia-surgery.org/news/2015/09/17/history-medicine-dr-roentgen-s-accidental-x-rays.

Encyclopædia Britannica, inc. (2021, March 23). *Wilhelm Conrad Röntgen*. Encyclopædia Britannica. https://www.britannica.com/biography/Wilhelm-Rontgen.

Frankel R. I. (1996). Centennial of Röntgen's discovery of x-rays. The Western journal of medicine, 164(6), 497–501

Jorgensen, T. J. (2017, October 11). *How Marie Curie Brought X-Ray Machines To the Battlefield*. Smithsonian.com. https://www.smithsonianmag.com/history/how-marie-curie-brought-x-ray-machines-to-battlefield-180965240/.

Khare, P., Nair, P., Khare, A., Singh, V., & Chatterjee, R. (2014). The Road To Radiation Protection: A Rocky Path. *Journal of Clinical and Diagnostic Research*, ZE01–ZE04. https://doi.org/10.7860/jcdr/2014/5832.5223

Michigan State University. (n.d.). *Thomas Alva Edison*. Thomas Alva Edison | Environmental Health & Safety | Michigan State University. https://ehs.msu.edu/lab-clinic/rad/hist-figures/edison.html.

Newman, T. (2018, January 9). *X-ray exposure: How safe are X-rays?* Medical News Today. https://www.medicalnewstoday.com/articles/219970#types.

Sansare, K., Khanna, V., & Karjodkar, F. (2011). Early victims of X-rays: a tribute and current    perception. Dento maxillo facial radiology, 40(2), 123–125. https://doi.org/10.1259/dmfr/73488299

Tretkoff, E. (n.d.). *This Month in Physics History*. American Physical Society. https://www.aps.org/publications/apsnews/200803/physicshistory.cfm.

University of Washington. (2018, November 15). *Featured History: Wilhelm Röntgen* . UW Radiology. https://rad.washington.edu/blog/featured-history-wilhelm-rontgen/.

UTMB Health. (n.d.). *The Discovery of the X-Ray* . Medical Discovery News . http://www.medicaldiscoverynews.com/shows/xray.html.

**Chapter 3:**

Banerjee, A. K., Beckmann, E., Busch, U., Buzzi, A., & Thomas, A. (2013). *The Story of Radiology* (Vol. 2). European Society of Radiology.

Benson, C. B., & Doubilet, P. M. (2014). The history of imaging in obstetrics. *Radiology*, *273*(2S), S92-S110.

Calabrese, E. J., & Dhawan, G. (2012). The role of x-rays in the treatment of gas gangrene: a historical assessment. *Dose-Response*, *10*(4), dose-response.

Dekaban, A. S. (1968). Abnormalities in children exposed to x-radiation during various stages of gestation: tentative timetable of radiation injury to the human fetus, part I. *Journal of Nuclear Medicine*, *9*(9), 471-477.

Farmelo, G. (1995). The discovery of X-rays. *Scientific American*, *273*(5), 86-91.

G. (n.d.). Achondroplasia. Retrieved May 09, 2021, from https://rarediseases.info.nih.gov/diseases/8173/achondroplasia#:~:text=Achondroplasia%20is%20a%20disorder%20of,small%20fingers%2C%20and%20normal%20intelligence

Halliday, A. (2017, September 15). Marie Curie invented mobile x-ray units to help Save wounded soldiers in World War I. Retrieved May 07, 2021, from https://www.openculture.com/2017/09/marie-curie-invented-mobile-x-ray-units-to-help-save-wounded-soldiers-in-world-war-i.html

Haygood, T. M. (1994). Chest screening and tuberculosis in the United States. *Radiographics*, *14*(5), 1151-1166.

Husen, L., Fulkerson, L. L., Del Vecchio, E., Zack, M. B., & Stein, E. (1971). Pulmonary tuberculosis with negative findings on chest x-ray films: a study of 40 cases. *Chest*, *60*(6), 540-542.

Martin, N. L. (2018, July 26). Radiology at Base hospital #28 in France During WW1. Retrieved May 09, 2021, from https://www.kumc.edu/wwi/base-hospital-28/clinical-services/radiology.html

Murphy Jr, W. A. (1990). Introduction to the history of musculoskeletal radiology. *RadioGraphics*, *10*(5), 915-943.

Quader, M. A., Sawmiller, C. J., & Sumpio, B. E. (2000). Radio contrast agents: history and evolution. In *Textbook of angiology* (pp. 775-783). Springer, New York, NY.

Strickland, D., & Stranges, A. N. (2018). X-rays: Laying the foundation of modern radiology, 1896-1930. *Medicina nei secoli*, *3*(2-3), 207-222.

**Chapter 4:**

Chen, H., Rogalski, M. M., & Anker, J. N. (2013). Advances in functional X-ray imaging techniques and contrast agents. *Physical Chemistry Chemical Physics: PCCP*, *14*(39), 13469–13486. https://doi.org/10.1039/c2cp41858d

Dammann, F., Bootz, F., Cohnen, M., Hassfeld, S., Tatagiba, M., & Kösling, S. (2014). Diagnostic imaging modalities in head and neck disease. *Deutsches Arzteblatt international*, *111*(23-24), 417–423. https://doi.org/10.3238/arztebl.2014.0417

Del Ciello, A., Franchi, P., Contegiacomo, A., Cicchetti, G., Bonomo, L., & Larici, A. R. (2017). Missed lung cancer: when, where, and why?. *Diagnostic and interventional radiology (Ankara, Turkey)*, *23*(2), 118–126.

https://doi.org/10.5152/dir.2016.16187

Hashmi, M. F., Katiyar, S., Keskar, A. G., Bokde, N. D., & Geem, Z. W. (2020). Efficient Pneumonia Detection in Chest Xray Images Using Deep Transfer Learning. *Diagnostics (Basel, Switzerland)*, *10*(6), 417. https://doi.org/10.3390/diagnostics10060417

Lusic, H., & Grinstaff, M. W. (2013). X-ray-computed tomography contrast agents. *Chemical reviews*, *113*(3), 1641–1666. https://doi.org/10.1021/cr200358s

Mahajan, R., & Singh, P. (2019). Radiocontrast Media: Applications and Concerns. *International journal of applied & basic medical research*, *9*(4), 191–192. https://doi.org/10.4103/ijabmr.IJABMR_322_19

Martínez Chamorro, E., Díez Tascón, A., Ibáñez Sanz, L., Ossaba Vélez, S., & Borruel Nacenta, S. (2021). Radiologic diagnosis of patients with COVID-19 [Diagnóstico radiológico del paciente con COVID-19]. *Radiologi´a*, *63*(1), 56–73. https://doi.org/10.1016/j.rxeng.2020.11.001

Nakashima, J., & Duong, H. (2021). Radiology, Image Production and Evaluation. *StatsPearl*.
https://www.ncbi.nlm.nih.gov/books/NBK553145/

Ngoya, P. S., Muhogora, W. E., & Pitcher, R. D. (2016). Defining the diagnostic divide: an analysis of registered radiological equipment resources in a low-income African country. *The Pan African medical journal*, *25*, 99. https://doi.org/10.11604/pamj.2016.25.99.9736

Rossi, F., Martinoli, C., Murialdo, G., Schenone, A., Grandis, M., Ferone, D., & Tagliafico, A. S. (2019). The primary role of radiological imaging in the diagnosis of rare musculoskeletal diseases. Emphasis on ultrasound. *Journal of ultrasonography*, *19*(78), 187–192. https://doi.org/10.15557/JoU.2019.0028

Satia, I., Bashagha, S., Bibi, A., Ahmed, R., Mellor, S., & Zaman, F. (2013). Assessing the accuracy and certainty in interpreting chest X-rays in the medical division. *Clinical medicine (London, England)*, *13*(4), 349–352. https://doi.org/10.7861/clinmedicine.13-4-349

Tafti, A., & Byerly, D. W. (2020). X-ray Image Acquisition. *StatsPearl*.
https://www.ncbi.nlm.nih.gov/books/NBK563236/

Wielpütz, M. O., Heußel, C. P., Herth, F. J., & Kauczor, H. U. (2014). Radiological diagnosis in lung disease: factoring treatment options into the choice of diagnostic modality. *Deutsches Arzteblatt international*, *111*(11), 181–187. https://doi.org/10.3238/arztebl.2014.0181

Zennaro, F., Oliveira Gomes, J. A., Casalino, A., Lonardi, M., Starc, M., Paoletti, P., Gobbo, D., Giusto, C., Not, T., & Lazzerini, M. (2013). Digital radiology to improve the quality of care in countries with limited resources: a feasibility study from Angola. *PloS one*, *8*(9), e73939. https://doi.org/10.1371/journal.pone.0073939

**Chapter 5:**

X-rays. (n.d.). National Institute of Biomedical Imaging and Bioengineering. Retrieved May 8, 2021, from https://www.nibib.nih.gov/science-education/science-topics/x-rays#:~:text=X%2Drays%20are%20a%20form,and%20structures%20inside%20the%20body.

Seibert, J. A. (2004). X-ray imaging physics for nuclear medicine technologists. Part 1: Basic principles of x-ray production. Journal of Nuclear Medicine Technology, 32(3), 139–147.

Carr, B.M., & Brown, G. (2014). Individual Project Shadow Puppetry Using the Kinect.

To X-ray or not to X-ray? (2016, April 14). World Health Organisation. https://www.who.int/news-room/feature-stories/detail/to-x-ray-or-not-to-x-ray-

Pogue, B. W., & Wilson, B. C. (2018). Optical and x-ray technology synergies enabling diagnostic and therapeutic applications in medicine. Journal of Biomedical Optics, 23(12), 121610. https://doi.org/10.1117/1.JBO.23.12.121610.

Thomas, J. W., LaMantia, C., & Magnuson, T. (1998). X-ray-induced mutations in mouse embryonic stem cells. Proceedings of the National Academy of Sciences, 95(3), 1114–1119. https://doi.org/10.1073/pnas.95.3.1114

Møystad, A., Svanaes, D. B., Risnes, S., Larheim, T. A., & Gröndahl, H. G. (1996). Detection of approximal caries with a storage phosphor system. A comparison of enhanced digital images with dental X-ray film. *Dentomaxillofacial Radiology, 25*(4), 202–206. https://doi.org/10.1259/dmfr.25.4.9084274

**Chapter 6:**

Achenbach, S. (November 2006). Computed Tomography Coronary Angiography. *Journal of the American College of Cardiology.* 48(10), 1919-1928. https://www.sciencedirect.com/science/article/pii/S0735109706020894?via%3Dihub

Ahmad, M., Pratz, G., Bazalova, M., Xing, L. (2014). X-Ray Luminescence and X-Ray Fluorescence Computed Tomography: New Molecular Imaging Modalities. *IEEE.* https://ieeexplore.ieee.org/document/6891106

Cademartiri, F., Runza, G., Belgrano, M., Luccichenti, G., Mollet, N.R., Malagutti, P., Silvestrini, M., Midiri, M., Cova, M., Mucelli, R.P., Krestin, G.P (2005). Introduction to Coronary Imaging with 64-Slice Computed Tomography. *La Radiologia medica.* 110, 16-41. https://www.researchgate.net/publication/7600137_Introduction_to_coronary_imaging_with_64-slice_computed_tomography

Chen, H., Rogalski, M.M., Anker, J.N. (2012). Advances in Functional X-ray Imaging Techniques and Contrast Agents. *Physical Chemistry Chemical Physics.* 14(39), 13469-13486. https://www.ncbi.nlm.nih.gov/pmc/articles/PMC3569739/

Coshocton Regional Medical Center. (2021). 64-Slice CT Scan. https://www.coshoctonhospital.org/services/diagnostic-services/64-slice-ct-scan/#:~:text=Like%20a%20traditional%20computed%20tomography,slices%E2%80%9D%20to%20produce%20actual%20pictures

Cramer, A., Hecla, J., Wu, D., Lai, X., Boers, T., Yang, K., Moulton, T., Kenyon, S., Arzoumanian, Z., Krull, W., Gendreau, K., Gupta, R. (2018) Stationary Computed Tomography for Space and other Resource-Constrained Environments. *Scientific Reports.* https://www.ncbi.nlm.nih.gov/

pmc/articles/PMC6155104/


Gingold, E. (2014). Modern Fluoroscopy Imaging Systems. *Image Wisely*. https://www.imagewisely.org/Imaging-Modalities/Fluoroscopy/Modern-Imaging-Systems


Johnson, T.R, Nikolaou, K., Busch, S., Leber, A.W., Becker, A., Wintersperger, B.J., Rist, C., Knez, A., Reiser, M.F., Becker, C.R. (2007). Diagnostic Accuracy of Dual-Source Computed Tomography in the Diagnosis of Coronary Artery Disease. *Invest Radiology*. 42(10), 684-691. https://pubmed.ncbi.nlm.nih.gov/17984765/


Johnson, T.R., Nikolaou, K., Becker, A., Leber, A.W., Rist, C., Wintersperger, B.J., Reiser, M.F., Becker, C.R. (2008). Dual-source CT for Chest Pain Assessment. *European Radiology*. 18(4), 773-780. https://www.ncbi.nlm.nih.gov/pmc/articles/PMC2270358/


Leber, A.W., Johnson, T., Becker, A., Ziegler, F., Tittus, J., Nikolaou, K., Reiser, M., Steinbeck, G., Becker, C.R., Knez, A. (2007). Diagnostic Accuracy of Dual-source Multi-Slice CT-Coronary Angiography in Patients with an Intermediate Pretest Likelihood for Coronary artery Disease. *European Heart Journal*. 28(19), 2354-2360. https://pubmed.ncbi.nlm.nih.gov/17644815/


Lee, D., Park, E.Y., Choi, S., Kim, H., Min, J., Lee, C., Kim, C. (2020). GPU-accelerated 3D Volumetric X-ray-Induced Acoustic Computed Tomography. *Biomedical Optics Express*. 11(2) 752-761. https://www.ncbi.nlm.nih.gov/pmc/articles/PMC7041460/


Li, Yang., Samant, P., Wang, S., Behrooz, A., Li, D., Xiang, L. (2020). 3-D X-Ray-Induced Acoustic Computed Tomography With a Spherical Array: A Simulation Study on Bone Imaging.


Lin, E and Alessio, A. (2009). What are the Basic Concepts of Temporal, Contrast, and Spatial Resolution in Cardiac CT? *Journal of Cardiovascular Computed Tomography*. 3(6), 403-408. https://www.ncbi.nlm.nih.gov/pmc/articles/PMC4752333/

Miller, J.C., Abbara, S., Mamuya, W.S., Thrall, J.H., Uppot, R.N. (2008). Dual-Source CT for Cardiac Imaging. *Journal of American College of Radiology.* 6(1), 65-68. https://www.jacr.org/article/S1546-1440(08)00456-0/fulltext

NIH. (2021). *National Institute of Biomedical Imaging and Bioengineering.* https://www.nibib.nih.gov/science-education/science-topics/mammography

Nikolaou, K., Flohr, T., Knez, A., Rist, C., Wintersperger, B., Johnson, T., Reiser, M.F., Becker, C.R. (2004). Advances in Cardiac CT Imaging: 64-Slice Scanner. *The International Journal of Cardiovascular Imaging.* 20(6), 535-540. https://pubmed.ncbi.nlm.nih.gov/15856639/

Petersilka, M., Bruder, H., Krauss, B., Stierstorfer, K., Flohr, T.G. (2008). Technical Principles of Dual Source CT. *European Journal of Radiology.* 68(3), 362-368. https://pubmed.ncbi.nlm.nih.gov/18842371/

Samant, P., Trevisi, L., Ji, X., Xiang, L. (2020) X-ray Induced Acoustic Computed Tomography. *Photoacoustics.* https://www.ncbi.nlm.nih.gov/pmc/articles/PMC7090367/

Tang, S., Yang, K., Chen, Y., Xiang, L. (2018). X-ray-induced Acoustic Computed Tomography for 3D Breast Imaging: A Simulation Study. *American Association of Physicists in Medicine.* https://aapm.onlinelibrary.wiley.com/doi/abs/10.1002/mp.12829

U.S. Food and Drug Administration. (2020). What is Computed Tomography? *FDA.* https://www.fda.gov/radiation-emitting-products/medical-x-ray-imaging/what-computed-tomography

**Chapter 7:**

X-Ray Pollution. Eco Dentistry Association. (n.d.). https://ecodentistry.org/green-dental-professionals/dental-office-waste/x-ray-pollution/.

Meo, S. A., Drees, A. M. A., Zadi, S. Z., Damgh, S. A., & Al-Tuwaijri, A. S. (2006). Hazards of X-Ray Radiation on the Quantitative and Phagocytic Functions of Polymorphonuclear Neutrophils in X-Ray Technicians. J Occup Health. 48(2):88-92. doi: 10.1539/joh.48.88.

Averbeck, D., Salomaa, S., Bouffler, S., Ottolenghi, a., Smyth, V., Sabatier, L. (2018). Progress in low dose health risl research: Novel effects and new concepts in low dose radiobiology.

Centers for Disease Control and Prevention. (2018). Health Problems Caused by Lead. Centers for Disease Control and Prevention. https://www.cdc.gov/niosh/topics/lead/health.html#:~:text=Exposure%20to%20high%20levels%20of,a%20developing%20baby's%20nervous%20system.

Nejaim, Y., Silva, A. I. V., Brasil, D. M., Vasconcelos, K. F., Haiter Neto, F., & Boscolo, F. N. (2015). Efficacy of lead foil for reducing doses in the head and neck: a simulation study using digital intraoral systems. Dento maxillo facial radiology.  Dentomaxillofac Radiol. 44(8): 20150065. doi: 10.1259/dmfr.20150065

Hewitt, P. J. (1993). Occupational health problems in processing of X-ray photographic films. Ann Occup Hyg.(3):287-95. doi: 10.1093/annhyg/37.3.287.

Lin, E. C. (2010). Radiation risk from medical imaging. Mayo Clinic proceedings. 85(12): 1142–1146. doi: 10.4065/mcp.2010.0260

Douple, E. B., Mabuchi, K., Cullings, H. M., Preston, D. L., Kodama, K., Shimizu, Y., Fujiwara, S., Shore, R. (2011). Long-term Radiation-Related Human Population: Lesson Learned from the Atomic Bomb survivors of Hiroshima and Nagasaki. Disaster Med Public Health Prep. 5(0 1): S122–S133. doi: 10.1001/dmp.2011.21

Komaroff, A. L. (2013). Ask the doctor: Should I worry about x-rays? Harvard Health. https://www.health.harvard.edu/staying-healthy/should-i-worry-about-x-rays.

**Chapter 8:**

Breedlove, S. M., & Watson, N. V. (2017). *Behavioral neuroscience* (8th ed.). Sinauer Associates, Inc., Publishers.

Mayo Clinic. (2020-a). *CT scan*. Retrieved May 7, 2021 from https://www.mayoclinic.org/tests-procedures/ct-scan/about/pac-20393675.

Mayo Clinic. (2019). *MRI*. Retrieved May 7, 2021, from https://www.mayoclinic.org/tests-procedures/mri/about/pac-20384768.
Mayo Clinic. (2020-b). *Positron emission tomography scan*. Retrieved May 7, 2021, from https://www.mayoclinic.org/tests-procedures/pet-scan/about/pac-20385078.

Mayo Clinic. (2020-c). *Ultrasound*. Retrieved May 7, 2021, from https://www.mayoclinic.org/tests-procedures/ultrasound/about/pac-20395177.

**Chapter 9:**

Andronikou, S. (2017). Letting go of what we believe about radiation and the risk of cancer in children. Pediatric Radiology. 47(1), 113-115. https://doi.org/10.1007/s00247-016-3697-5

Baumann, C., Singmann, H., Gershmann, S. J., & von Helversen, B. (2020). A linear threshold model for optimal stopping behaviour. PNAS. 117(23), 12751-12755. https://doi.org/10.1073/pnas.2002312117

Bertell, R. (1977) X-ray exposure and premature aging. Journal of Surgical Oncology. 9(4), 379-391. https://doi.org/10.1002/jso.2930090409

Chou, W. S., & Klein, E. M. (2018). Addressing health-related misinformation on social media. JAMA, 320(23), 2417-2418. https://doi.org/10.1001/jama.2018.16865

Cohen-Kerem, R., Nulman, I., Abramox-Newerly, M., Medina, D., Maze, R., Brent, R. L., & Koren, G. (2006) Diagnostic radiation in pregnancy: perception versus true risks. Journal of Obstetrics and Gynaecology Canada, 28(1), 43-48.https://doi.org/10.1016/S1701-2163(16)32039-4

de Gonzalez, A. B., & Darby, S. (2004). Risk of cancer from diagnostic X-rays: estimates for the UK and 14 other countries. Lancet. 363(9406), 345-351. https://doi.org/10.1016/S0140-6736(04)15433-0

De Santis, M., Di Gianantonio, E., Straface, G., Cavaliere, A. F., Caruso, A., Schiavon, F., Berletti, R., & Clementi, M. (2005). Ionizing radiations in pregnancy and teratogenesis: a review of literature. *Reproductive Toxi-*

*cology.* 20(3), 323-329. https://doi.org/10.1016/j.reprotox.2005.04.004

Doss, M. (2018). Are we approaching the end of the linear no-threshold era? Journal of Nuclear Medicine. 59(12), 1786-1793. https://doi.org/10.2967/jnumed.118.217182

Fazio, L. K., Brashier, N. M., Payne, B. K., & Marsh, E. J. (2015). Knowledge does not protect against illusory truth. Journal of Experimental Psychology: General, 144(5), 993–1002. https://doi.org/10.1037/xge0000098

Feinendegen, L. E., Pollycove, M., & Sondhaus, C. A. (2004). Responses to low doses of ionizing radiation in biological systems. Nonlinearity in Biology, Toxicology, and Medicine. 2(3), 143-171. https://doi.org/10.1080/15401420490507431

Goertzel, T. (2010). Conspiracy theories that target specific research can have serious consequences for public health and environmental policies. EMBO reports, 11(7), 493-499. https://doi.org/10.1038/embor.2010.84

Harvard Health Publishing. (2020, January 29). Radiation risk from medical imaging. https://www.health.harvard.edu/cancer/radiation-risk-from-medical-imaging

Ratnapalan, S., Bentur, Y., & Koren, G. (2008). "Doctor, will that x-ray harm my unborn child?". CMAJ. 179(12), 1293-1296. https://doi.org/10.1503/cmaj.080247

Stojanov, A., & Halberstadt, J. (2020) Does lack of control lead to conspiracy beliefs? A meta-analysis. European Journal of Social Psychology. 50(5), 955–968. https://doi.org/10.1002/ejsp.2690

Tobah, Y. B. (2020, March 07). Is it safe to have an X-ray during pregnancy? Mayo Clinic. https://www.mayoclinic.org/healthy-lifestyle/pregnancy-week-by-week/expert-answers/x-ray-during-pregnancy/faq-20058264#:~:text=The%20possibility%20of%20an%20X,pot ential%20risk%20to%20a%20baby

Tomà, P., Cannatà, V., Genovese, E., Magistrelli, A., & Granata, C. (2017). Radiation exposure in diagnostic imaging: wisdom and prudence, but still a lot to understand. Radiologia Medica, 122(3), 215-220 https://doi.org/10.1007/s11547-016-0709-3

Vandergriendt, C. (2020, March 15). The Dunning-Kruger Effect Explained. Healthline. https://www.healthline.com/health/dunning-kruger-effect


**Chapter 10:**

ACR. (2021, January 13). *NCRP Recommends Against Routine Gonadal Shielding.* ACR. https://www.acr.org/Media-Center/ACR-News-Releases/2021/NCRP-Recommends-Against-Routine-Gonadal-Shielding

AAPM. (2019). *AAPM Position Statement on the Use of Patient Gonadal and Fetal Shielding.* AAPM. https://www.aapm.org/org/policies/details.asp?id=468&type=PP&current=true

Balbi, M., Caroli, A., Corsi, A., Milanese, G., Surace, A., Marco, F. D., Novelli, L., Silva, M., Lorini, F. L., Duca, A., Cosentini, R., Sverzellati, N., Bonaffini, P. A. & Sironi, S. (2020) Chest X-ray for predicting mortality and the need for ventilatory support in COVID-19 patients presenting to the emergency department. *Eur Radiol 31*, 1999–2012. https://doi.org/10.1007/s00330-020-07270-1

Bjarnason, T.A., Rees, R., Kainz, J., Le, L.H., Stewart, E.E., Preston, B., Elbakri, I., Fife, I.A.J., Lee, T.-Y., Gagnon, I.M.B., Arsenault, C., Therrien, P., Kendall, E., Tonkopi, E., Cottreau, M. & Aldrich, J.E. (2020). COMP Report: A survey of radiation safety regulations for medical imaging x-ray equipment in Canada. *J Appl Clin Med Phys, 21*: 10-19. https://doi.org/10.1002/acm2.12708

Borkowski, A. A., Viswanadhan, N. A., Thomas, L. B., Guzman, R. D., Deland, L. A., & Mastorides, S.M. (2020). Using Artificial Intelligence for COVID-19 Chest X-ray Diagnosis. *Federal practitioner : for the health care professionals of the VA, DoD, and PHS, 37(9)*, 398–404. https://doi.org/10.12788/fp.0045

Business Wire. (2020, September 30). *KA imaging's portable dual-energy x-ray receives Canadian medical device licence from Health Canada.* Business Wire. https://www.businesswire.com/news/home/20200930005044/en/

Liang, C., Zhang, S., Cheng, L., Xie, J., Zhai, F., He, Y., Wang, Y., Chai, Z., & Wang, S. (2020). Thermoplastic membranes incorporating semiconductive metal–organic frameworks: An advance on flexible X-ray detectors. *Angewandte Chemie International Edition, 59(29)*, 11856-11860. https://doi.org/10.1002/anie.202004006

Dumé, I. (2021, March 31). *Flexible detector takes high-resolution X-ray images in 3D.* physicsworld. https://physicsworld.com/a/flexible-detector-takes-high-resolution-x-ray-images-in-3d/

Florida State University. (2020, August 31). New X-ray detection technology developed. ScienceDaily. Retrieved May 8, 2021 from www.sciencedaily.com/releases/2020/08/200831154359.htm

Gall, A.L. (2020, November 18). *New 3D colour X-rays made possible with CERN technology.* CERN. https://home.cern/news/news/knowledge-sharing/new-3d-colour-x-rays-made-possible-cern-technology

Gange, C. P., Pahade, J. K., Cortopassi, I., Bader, A. S., Bokhari, J., Hoerner, M., Thomas, K. M., Rubinowitz, A. M. (2020). Social distancing with portable chest radiographs during the COVID-19 pandemic: Assessment of radiograph technique and image quality obtained at 6 feet and through glass. *Radiography: Cardiothoracic Imaging, 2(6).* https://doi.org/10.1148/ryct.2020200420

Harvard Health Publishing. (2020, January 29). *Radiation risk from medical imaging.* Harvard Health Publishing. https://www.health.harvard.edu/cancer/radiation-risk-from-medical-imaging

Health Quality Ontario. (2016). *Report and recommendations on modernizing Ontario's radiation protection legislation.* Health Quality Ontario. https://www.hqontario.ca/Portals/0/documents/health-quality/modernizing-ontario-radiation-protection-legislation-report-en.pdf

Ippolito, D., Pecorelli, A., Maino, C., Capodaglio, C., Mariani, I., Giandola, T., Gandola, D., Bianco, I., Ragusi, M., Franzesi, C. T., Corso, R., & Sironi, S. (2020). Diagnostic impact of bedside chest X-ray features of 2019 novel coronavirus in the routine admission at the emergency department: case series from Lombardy region. *European Journal of Radiology, 129*, 109092. https://doi.org/10.1016/j.ejrad.2020.109092

Jacobi, A., Chung, M., Bernheim, A., & Eber, C. (2020). Portable chest

X-ray in coronavirus disease-19 (COVID-19): A pictorial review. *Clinical imaging, 64,* 35–42. https://doi.org/10.1016/j.clinimag.2020.04.001

Jenkins, H. J., Downie, A. S., Moore, C. S., & French, S. D. (2018). Current evidence for spinal X-ray use in the chiropractic profession: a narrative review. *Chiropractic & manual therapies, 26,* 48. https://doi.org/10.1186/s12998-018-0217-8

Kasten Y., Doktofsky D., & Kovler I. (2020) End-To-End Convolutional Neural Network for 3D Reconstruction of Knee Bones from Bi-planar X-Ray Images. In: Deeba F., Johnson P., Würfl T., Ye J.C. (eds) Machine Learning for Medical Image Reconstruction. *MLMIR 2020. Lecture Notes in Computer Science, vol 12450.* Springer, Cham. https://doi.org/10.1007/978-3-030-61598-7_12

Mozdy, M. (2020, April 8). *Safer, PPE-conserving X-Rays for patients at university hospital.* University of Utah School of Medicine. https://medicine.utah.edu/radiology/news/2020/04/x-ray-through-glass.php

Mun, S. K., Wong, K. H., Lo, S. B., Li, Y., & Bayarsaikhan, S. (2021). Artificial intelligence for the future radiology diagnostic service. *Frontiers in molecular biosciences, 7,* 512. http://doi.org/10.3389/fmolb.2020.614258

NCRP. (2021). *NCRP Statement No. 13: NCRP recommendations for ending routine gonadal shielding during abdominal and pelvic radiography.*National Council on Radiation Protection and Measurements. https://ncrponline.org/wp-content/themes/ncrp/PDFs/Statement13.pdf

Oakley, P. A., & Harrison, D. E. (2021). Are Continued Efforts to Reduce Radiation Exposures from X-Rays Warranted?. *Dose-response : a publication of International Hormesis Society, 19(1),* 1559325821995653. https://doi.org/10.1177/1559325821995653

Oakley, P. A., & Harrison, D. E. (2020). Death of the ALARA Radiation Protection Principle as Used in the Medical Sector. *Dose-response : a publication of International Hormesis Society, 18(2)*, 1559325820921641. https://doi.org/10.1177/1559325820921641

O'Connor, M., Stowe, J., Potocnik, J., Giannotti, N., Murphy, S., & Rainford, L. (2021). 3D virtual reality simulation in radiography education: The students' experience. *Radiography (London, England : 1995), 27(1),*

208–214. https://doi.org/10.1016/j.radi.2020.07.017

Syeda-Mahmood, T. (2020, November 4). *IBM AI algorithms can read chest X-rays at resident radiologist levels.* IBM Research Blog. https://www.ibm.com/blogs/research/2020/11/ai-x-rays-for-radiologists/

Sykes P. J. (2020). Until There Is a Resolution of the Pro-LNT/Anti-LNT Debate, We Should Head Toward a More Sensible Graded Approach for Protection From Low-Dose Ionizing Radiation. *Dose-response : a publication of International Hormesis Society, 18(2)*, 1559325820921651. https://doi.org/10.1177/1559325820921651

The Fifth International Symposium on The System of Radiological Protection. (2020). *Annals of the ICRP, 49*(1_suppl), 5–8. https://doi.org/10.1177/0146645320959792

Toppenberg, M., Christiansen, T. E. M., Rasmussen, F., Nielsen, C. P., & Damsgaard, E. M. (2020) Mobile X-ray outside the hospital: a scoping review. *BMC Health Serv Res, 20*, 767. https://doi.org/10.1186/s12913-020-05564-0

University of Maryland Baltimore County. (2019, November 7). *New X-ray technology could revolutionize how doctors identify abnormalities.* ScienceDaily.Retrieved May 8, 2021 from www.sciencedaily.com/releases/2019/11/191107160601.htm

Xu, LJ., Lin, X., He, Q., Worku, M., & Ma, B. (2020) Highly efficient eco-friendly X-ray scintillators based on an organic manganese halide. *Nat Commun 11*, 4329 (2020). https://doi.org/10.1038/s41467-020-18119-y